Precautionary Risk Management

Precautionary Risk Management

Dealing with Catastrophic Loss Potentials in Business, the Community and Society

Mark Jablonowski

palgrave
macmillan

First published 2006 by
PALGRAVE MACMILLAN
Houndmills, Basingstoke, Hampshire RG21 6XS and
175 Fifth Avenue, New York, N.Y. 10010
Companies and representatives throughout the world

PALGRAVE MACMILLAN is the global academic imprint of the Palgrave Macmillan division of St. Martin's Press, LLC and of Palgrave Macmillan Ltd. Macmillan® is a registered trademark in the United States, United Kingdom and other countries. Palgrave is a registered trademark in the European Union and other countries.

ISBN 13: 978–0–230–01352–0 hardback
ISBN 10: 0–230–01352–X hardback

This book is printed on paper suitable for recycling and made from fully managed and sustained forest sources.

A catalogue record for this book is available from the British Library.

Library of Congress Cataloging-in-Publication Data

Jablonowski, Mark, 1955–
Precautionary risk management : dealing with catastrophic loss potentials in business, the community and society / by Mark Jablonowski.
p. cm.
Includes bibliographical references and index.
ISBN 0–230–01352–X (cloth)
1. Risk management. 2. Emergency management. I. Title.

HD61.J32 2006
658.15'5–dc22 2006046010

10 9 8 7 6 5 4 3 2 1
15 14 13 12 11 10 09 08 07 06

Printed and bound in Great Britain by
Antony Rowe Ltd, Chippenham and Eastbourne

Contents

List of Tables and Figures

Tables

Figures

Preface: A World of Hazards

Today, the world in which we live brings to our attention the importance of high-stakes decisions in the management of risk. The increasing complexity of modern systems, be they physical, economic or social, brings with it the threat of increased risks. For a variety of reasons, the threats we face have potential impacts greater than ever before in history. Some of the threats themselves are indeed "man made", such as the disposal of industrial waste, and political and social instabilities that promote terrorism and warfare. Other, age old physical perils are exacerbated by industrial and social progress, such as the development of areas prone to severe wind and earthquake damage. Given its increasing importance, we need to carefully *re-examine* the question: How do we deal with exposures that can have tremendous, possibly catastrophic, negative impacts, on an individual, organizational and societal level?

While we face a variety of risks in the world today, we will focus here on what has generally been termed *hazard risk*. This form of risk arises from physical properties of the world, and manifests its effects directly on *nature*. Nature includes people and their property. This is in contrast to the study of economic risks, which affect primarily financial variables. Of course, hazard risk may ultimately have financial effects. The ruin of a business enterprise is defined in terms of financial variables, even though physical perils in the form of hazard risk may represent the "root cause".

Hazard risks can themselves be classified as natural and man-made. Natural risks include the age-old perils of fire, wind, earthquake and flood. The modern world introduces a variety of perils produced by man. These include the risks associated with technological progress, including the failure of technology to function as intended, resulting in physical harm to humans and their environment. Technological byproducts include waste production, in the form of pollutants. Sometimes hazard risk is extended to perils that include the intentional acts of men, for political, criminal or even psychological reasons. Terrorism and war are among them. While these perils may to some extent be treated as purely physical, their roots are political. Complete analysis of these risks therefore requires use to go beyond hazard risk into the assessment of political risk.

We choose this focus on hazard risk for two reasons. First, and most importantly, hazard risks clearly have the most impact on the preservation and continuation of life as we know it. Financial hardships can have serious impacts, some of them manifested physically. The Great Depression in the United States was a case of economic catastrophe, with widespread negative results. Despite its financial magnitude, and its ultimate effects on the quality of life, economic recession, and even depression, cannot be considered threatening to life as we know it. On the other hand, the unmitigated disbursement of hazardous pollutants can.

On the individual level, people suffer financial hardship, and survive. On the business level, unmitigated hazard risk is often the cause of catastrophic financial ruin. Societal financial crisis affects the quality of life, but not its continuation. On the whole, nature remains fairly impervious to economic risk. With respect to individual businesses, hazard risk also remains a leading cause of financial failure in the world today.

We also concentrate on hazard risk here so as to limit the topic of risk to manageable proportions. In doing so, we must keep in mind that many of the techniques discussed here with respect to hazard risk apply to economic risk as well. The *safety-first* principle in the management of the risk of business investments requires catastrophic (ruin) potential be assessed first and foremost. Investments that violate the safety criteria (in terms of this ruin threshold) should be rejected, regardless of their "upside" potential. Safety-first investment strategies are based on precautionary risk management principles. Economic legislation and regulation often proceeds on a precautionary basis as well (for example, US antitrust regulation). The application of risk management principles to macroeconomic activity is a vast area for research. Suffice it to say, these challenges are outside of the scope, and capability, of any single text. This one remains, therefore, focused on hazard risk. The hope is that the ideas can be applied to economic risks with similar benefits.

The reader is forewarned that we make no firm conclusions here on the "right" way to deal with high-stakes risk. We do narrow the options, however, to precautionary avoidance (including effective loss prevention), and voluntary acceptance of risk, or what we call "fatalism". Statistical extensions of the basic principles of economic optimization, based on expected value cost/benefit analysis, are shown to be inapplicable in the high-stakes arena. We further elucidate the fundamentals of precaution, based on the notion of *minimax*, illuminating as reasonably as possible both its strengths and weaknesses. The stakes are indeed high enough that simplistic pronouncements, such as precaution is "too expensive" or "impractical", must be thoroughly examined. So must the interests of those involved in the debate, as impartiality with regard to risks is hard to come by. As we will point out (often), when risk decisions are made, especially high-stakes ones, there may be significant winners and losers, in terms of both wealth, health and even existence. We have to try to get by special interests if we are to come to any realistic conclusions with respect to high-stakes risks. So, our goal in the chapters that follow is to try to represent the fundamentals of *precautionary risk management* as objectively, and succinctly, as possible.

Acknowledgements

Portions of this work have appeared in articles previously published in *The John Liner Review: The Quarterly Journal of Advanced Risk Management Strategies, Risk Management, The Journal of the Society of Chartered/Property Casualty Underwriters*. Materials in Chapter 2, Section 2.1, "The 'Iceberg' Model of Risk", originally appeared in "Managing the Iceberg Model of Risk" published in the July 2005 issue of the *CPCU eJournal*, located at www.cpcusociety.org, and is reprinted with permission of the CPCU Society, Malvern, PA. Sections of Chapter 2, "The ABC's of High-Stakes Risk", and Appendix A, "A Working Model of Precaution", appeared in the *John Liner Review*, and are reprinted here with the kind permission of Standard Publishing Corporation, Boston, Massachusetts. Quoted throughout, are sections of "Facing Risk in the 21st Century", Copyright 2004, *Risk Management* and the Risk and Insurance Management Society, Inc. (all rights reserved). The author would like to thank these publications for their continuous support of ideas that link formal analysis with intuitive applications.

1 Statistical Risk and its Treatment

A great deal of day-to-day risk management, be it on a personal, business or community level, proceeds "statistically". That is, we attempt to make sense of random or chance phenomena using averages and other representations that allow us to make reasonable decisions over time. We will see that much of what we call "risk" is amenable to statistical treatment in terms of aggregates over time. This allows us to identify the best, or optimal, course of action with respect to these risks using basic economic theory. The same principles we apply to rational optimization, in business and in life in general, apply to statistical risks. The statistical nature of many risks also permits verification of our results over time. Statistical analysis does have boundaries of applicability, however. We must be very careful not to extend statistical analysis beyond these boundaries. To understand, and manage, the full spectrum of risks we deal with, we must understand both the strengths and the very real limitations of the statistical approach.

1.1 Economic optimization and cost/benefit analysis

Economics is the study of allocation of scarce resources so as to best satisfy humans' wants and needs. In seeking the "best", we must always balance rewards, or benefits, against costs. Costs represent the commitment of scarce resources. The question behind the economic analysis of costs and benefits becomes: How do we achieve the maximum net benefit (i.e., benefits less costs)? In other words, economic choice is about *optimization*.

In many cases the application of cost/benefit analysis is straightforward. In the simplest case, we engage in some activity as long (and to the point where) benefits exceed cost. As long as some activity entails some positive contribution to overall satisfaction (however we happen to measure that), we engage in the activity. That activity could be eating hot dogs, investing in a stock or choosing a home. On the wider, social scale cost/benefit analysis is applied to a variety of important decisions that affect human life.

When cost and benefits can be precisely measured, we say that the analysis is *deterministic*. If it costs, in monetary terms, $100,000 to produce 100 widgets, and we can sell those widgets for $110,000, it makes economic sense to do so. Our net gain, or profit, is $10,000. We have gained a level of satisfaction (here somewhat simplistically measured in dollars and cents) greater than the expenditure needed to achieve such satisfaction. Resources in this case were used efficiently.

By consistently following this rule, we optimize our overall use of resources to provide maximal gains, not just for the individual, but for society as well. Of course complications can arise along the way. When benefits and costs are incurred over some period of time, we need to properly account for income streams in terms of the time value of money (i.e., the nominal dollar is "worth" morc today than a dollar received a year from now). Effective cost/benefit requires that the results of the optimization be properly discounted for benefits received in the future. Debates also surface on the proper means of *measuring* costs and benefits. All things considered, cost/benefit analysis still provides fruitful guidance on important decisions.

1.2 The nature of risk

We can not always determine our costs and benefits exactly. Different types of *uncertainty* exist in the real-world of application. The idea of risk is based on the concept of uncertainty due to randomness. We need to take this uncertainty into consideration when making real-world decisions about risk.

Risk is a concept that is defined and used in many ways. Sometimes we use the word risk as a synonym for danger or threat of harm. More generally, risk is used to refer to any event that entails some likelihood of loss or damage. While definitions vary, all concepts of risk entail the idea of the chance occurrence of some untoward event.

While risk always implies threat, damage or harm, we cannot know exactly where or when risky events will occur. The best we can do is get some idea of their long-run relative frequency. We call this long-run frequency of occurrence *probability*. The probability of event x, p(x), can be defined as,

$$p_x = \frac{\text{Number of outcomes in which x occurs}}{\text{Total number of outcomes}} \qquad \text{(Equation 1.1)}$$

We can visualize the concept of probability using a simple urn, or bowl, model as shown in Figure 1.1. Picture a bowl filled with ten colored balls, the selection of any one representing an "outcome" of our experiment. Nine of the balls are white, and one is black. We mix the ball thoroughly and draw a ball without looking (as ensuring "randomness" in our pick). The probability of drawing a black ball (event X) is, by our formula above, 1/10 or .1. In repeated draws, we expect to draw a black ball one time out of ten (though our actual outcomes, again, will always be unknown prior to our experiment). This probability number gives us some idea of what proportion of times we can expect to draw the black ball over a large number of draws (the "long-run").

The accidental losses we deal with in risk management are random events. They are neither expected nor intended from the standpoint of the entity attempting to manage them. As such, they can be modeled using our bowl of colored balls. Assume for example, that drawing a black ball represents the occurrence of some event, while a white ball represents non-occurrence. The bowl experiment can now serve as a model of, say, the probability that we will have an auto accident over the next year. Black balls are labeled "accident" and white balls as "no accident". By varying the proportion of black balls to white balls, we can model any probability we desire.

In the assessment and analysis of risk, we associate with some accidental event x untoward consequences that we assume, for now, can

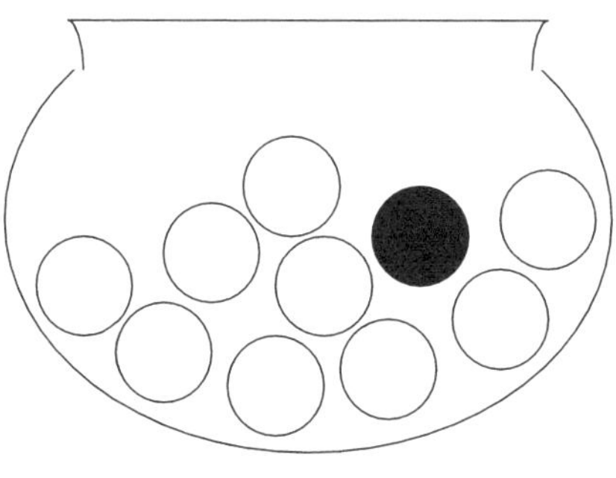

Figure 1.1 The "Urn Model" of Probability

be measured in monetary terms. So for each draw of a black ball from our urn of colored balls, we assume that an accident takes place and that accident is associated with some negative monetary value. Just how much negative monetary value is usually determined by a host of physical properties, and their interactions. Once again, these events, and hence the associated monetary losses occur randomly. We don't know when the next one will occur. We can only get some idea of their *average* value over time. When the loss amount associated with some event ($X) is fixed, we can calculate the average as follows:

$$\frac{\text{Number of outcomes in which event x occurs} \bullet \$X}{\text{Total number of outcomes}} \qquad \text{(Equation 1.2)}$$

Where the symbol "•" denotes multiplication, and the "bar", division. So, if the only possible loss amount is fixed at $10,000, the total number of outcomes in which x occurs is one, and the total number of possible (or observed) occurrences is ten, we say the average loss is $1,000. Going forward in time, we find that in repeated sequences of losses (draws from an urn, years), we observe instances where no losses are observed, and the occasional (random) $10,000 loss. If we add up the number of losses, and then divide them by the number of observations, the number will equal $1,000 over time. As we make observations over a large number of samples, or alternatively, years, the observed average more closely matches the "true" average. The amount of error is based on the theory of sampling error.

We could re-write Equation 1.2 as:

$$\left[\frac{\text{Number of outcomes in which events x occurs}}{\text{Total number of outcomes}}\right] \bullet \$X \qquad \text{(Equation 1.3)}$$

However, the bracketed term is simply our definition of probability (Equation 1.1), so the average can also be calculated as,

$$P_x \bullet \$X \qquad \text{(Equation 1.4)}$$

This average is known as the *expected value* (EV) of loss.

1.3 Using expected values

Through time, actual outcomes converge to the expected value. Say we draw repeatedly from our bowl of colored balls. The proportion of black to white is one in ten, so the probability of drawing a black ball is .1. Drawing a black ball indicates an "accident" costing $10,000. We draw from the bowl, each draw constituting a "trial". A typical outcome is shown in Figure 1.2. Note that over a number of trials, the results "average out" to $1,000 of losses per trial, as specified by our theoretical formula. If, as is often the case with natural phenomenon and other hazard risks, the averages are measured over time, say a one year interval, then the "trials" can be viewed as observed "years". Statistical results in these cases average out over a number of years of observations.

When multiple loss amounts are considered (or, more realistically, losses vary continuously over the scale of possible loss amounts), we can associate with each of them a probability. Doing so, we define a *probability distribution* of loss. The average of a probability distribution is found by multiplying all possible loss amounts by their associated probabilities, and adding them together. These two components, the relative frequency (probability) and amount of loss (often referred to as its severity) define the characteristics of any exposure to accidental loss. We can also often define probability distributions using mathematical equations. While we will focus on point estimates here, to make things more clear, the results extend to probability distributions as well.

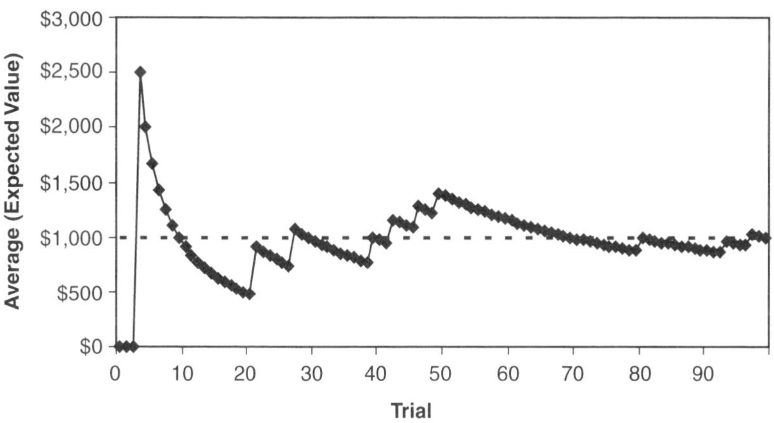

Figure 1.2 Averages Converge Over a Number of Trials

To represent statistical risk in analysis we use averages. Averages tell us how much statistical risk we face, and help us determine the best way to reduce it or eliminate it. Using expected values we can optimize the amount of risk management activity (loss prevention, loss control, avoidance, etc.) we utilize based on cost/benefit comparisons. Random inputs are treated just like any other economic activity. We simply add the qualifier "on average" to account for the random nature of losses, knowing that over time the optimal result will be realized. Obviously, to make expected values useful in this regard, we need to be able to experience a suitable amount of time (or, number of trials) within which the averages can work themselves out (i.e., stabilize). We'll keep this condition in mind when we turn our attention to the treatment of high-stakes, catastrophic risks.

1.4 Determining probabilities from data

To use expected values to manage risk we also need to know the probabilities associated with accidental losses. We have assumed in the discussion so far that we know the composition of our urn before hand. We don't know what the next draw will be, but we do know the mix of balls, and from that we can compute expected values. In the real-world, we rarely get to see the "big picture" in terms of total number of occurrences, and the number of occurrences we can expect some event x to occur. That is, we don't usually deal in situations where we can completely enumerate the possible outcomes, as in the case of our simple urn example. We have to estimate the probabilities themselves, through the process of *sampling*. Sampling involves observing individual instances from a population of possible outcomes. In our urn example, the *population* is the ten balls in the bowl. More often, we are unsure of the size of the population, and may even consider it infinite. Where sampling is controlled, we can take samples under controlled conditions. For example, if we do not know the composition of our bowl before hand, and the bowl is opaque, we can draw individual balls and note their color. We infer the true composition of the bowl by repeated sampling. Usually, in the real-world, we cannot control the process. We observe samples of events as they happen through time (auto accidents, for example).

Obviously, the more we sample (that is, the greater the sample size), the more confident we can become in the result. For example, if we draw three balls from a bowl full of ten balls, mixed in the proportion nine white balls and one black, we will not be able to infer the mix of balls as well as if we drew 6, 10, or 20 times. We assume here that each time we are sampling "with replacement", meaning the balls are returned to the bowl after every draw (so as to preserve a uniform sample size). A statistical analysis will permit us to identify the exact probability that the bowl's composition is anywhere from 0 to 10 black balls. We can do so mathematically, by specifying the total number of combinations of balls possible for any given sample size. If we assume drawing each group of a given sample size proceeds with equal probability, we can calculate the probability we would get our specific result, given the various possible population values of 0 to 10 black balls. To do so, we use the mathematical formula for calculating what are called Binomial ("two outcome") probabilities. Say that in our 3-ball sample, all three draws are white. Applying the formula we find, not surprisingly that the "most probable" composition of the bowl is all white balls (i.e., probability of a black ball = 0). However, due to the small sample size, the probability of at least one black ball being in the mix is still very high. High enough that additional sampling is warranted if we want to make pronouncements about the composition of the bowl with a strong degree of confidence.

Table 1.1 shows sample sizes required to gain specified degrees of confidence in our results. Again, as sample size increases, so does our confidence in the result. Note that sample sizes need to be rather large to gain a high degree of confidence, even in simple sampling situations. Often, sampling error is expressed as an interval around the average value. These statistical error bounds are known as *confidence intervals*.

Repeated sampling under controlled conditions plays a large part in the overall process of risk management. To test the integrity of a safety critical bolt we may test hundreds or even thousands of samples. However, this type of controlled sampling can rarely be achieved with

Table 1.1 Sample Size and Estimation Error

Sample Size	Percent Error
5	22
10	16
20	10
50	7
100	5
250	3
500	2
1000	1

accidental phenomenon that occurs in the real environment of risk. For example, the geological data on earthquakes in any part of the world is generally fairly limited. This means the sampling uncertainty about probability estimates will be great. What's more, the mathematical theory of sampling error makes some strong assumptions about underlying populations and the process itself. Chief among those is that the population is relatively stable throughout the sampling process. Stability is far from guaranteed in the real-world of application. When sampling can proceed on the basis that all or most of its underlying assumptions are valid, it provides us with average values and probability distributions that we can be relatively confident in, over time. At least theoretically, sampled probabilities provide us with a sound guide for action.

We can actually update our expected value decisions based on sampled data as we acquire it over time. To do so we use so-called *Bayesian* statistical methods. Applying *Bayes Theorem*, we methodically incorporate new data into our expected value analysis, as it becomes available. The credibility of our results, however, is still bound by sample size.

1.5 Extending the credibility of statistical results

Observations on the credibility of data are based on a simple averaging of statistical data, itself based on observations of actual losses over time. For individual exposures, say the average business entity, this "sampling" is based on experienced losses over some available time period. We have already suggested that statistical sampling under controlled conditions can only be carried out in a very limited arena. Other ways of extending credibility of sampled results include expanding the statistical base by looking at a wider group of similar exposures (as sort of "semi-controlled" sampling), and using logical methods based on event trees and fault trees.

Credibility may be expanded somewhat by considering a wider statistical base of information (i.e., "industry data", or "national" or event "world-wide data"). This allows many more data points to enter the analysis. I may have 10 years of data for one firm in the industry, but if I have 10 such firms I have extended my database to 100. The problem is that as we base the comparison on a wider group, the similarity of conditions to our own decreases, and the likelihood estimates that result are less credible. For example, from the standpoint of an individual firm's loss data, the wider the industry base, the less similar the industry will be to ours. The same goes for time series comparisons, i.e., data over time.

The individual company, and indeed fundamental underlying process that cause losses, change over time. The farther we go into the past to collect data, the less relevant that data is to today's operations. The uncertainty that results is not something that can itself be treated statistically. It is a form of knowledge imperfection that suggests various degrees of possibility, those possibilities expanding the more dissimilar our benchmark firms, and the farther we go back in time.

In some cases, such as natural phenomenon, like earthquakes, windstorms and floods, the exposure is so widespread as to make even an expanded database fairly limited. Also, the physical conditions may be so unique as to make any multiple comparisons invalid. For example, we would not be reasonable to combine the experience of Turkey and California to increase the credibility of statistical prediction of earthquakes for either location. Certain similarities in physical features may make this possible, but these combined estimates would have to be treated with extreme care. They would only be as good as the underlying scientific analogies permit. Unfortunately, there are no formal methods to judge the credibility of statistical reasoning by analogy.

Another way to expand the credibility of limited statistical data is the use of engineering techniques such as *event trees*. Event trees use more readily available "sub-event" data, such as the failure of a sprinkler system, logically combine them using the theory of probability, and come up with estimates of low probability outcomes for which statistical data is simply not available.

Figure 1.3 shows a simple event tree. Here we work from initiating event, to final outcomes, where the initiating event is something that "starts" the loss process. The probability for initiating events is usually determined statistically. These events usually occur with sufficient frequency so as to make their statistical probabilities fairly accurate. The example we use here is an event tree (greatly simplified) for a fire in a large manufacturing plant. Initiating events for fire include things like careless smoking, electrical malfunctions or acts of God, such as a lightning strike. The initiating event probability is computed on an annual basis. Here, company and related data suggest the fire probability is once every five years, or .2. Once a fire starts, we use engineering and process knowledge to identify the possible events that follow. In our case, the next step is detection. We use a simple dichotomy to identify the possibilities: Either the fire is detected (and promptly extinguished) or it is not. At this point, probabilities are assigned to the "branches" of the tree. Based on our own individual experience, and information obtained from our fire insurer and other sources, we identify the probability of

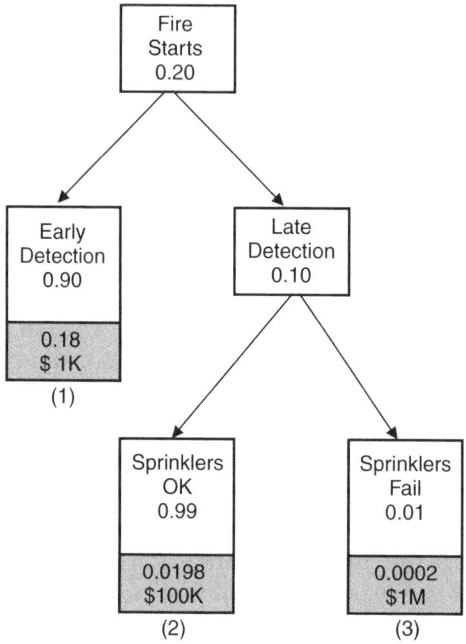

Figure 1.3 A Simple Event Tree

detection as .9. The probability of an undetected fire is just the compliment, 1 minus .9, or .1. Notice that we have one terminal, or "outcome" branch, consisting of the sequence, fire starts AND it's promptly detected and extinguished. We include in the outcome a possible loss amount. Rapid detection implies limited loss, in this case $1,000. We can also calculate the probability of this outcome, using the theory of probability and the "sub event" probabilities. The "AND" conjunction means that we multiply the probabilities of the sub-events to get the outcome probability: Fires start (.2) x detection (.9) = .18. This is the probability of a small ($10,000) fire. Multiplying the probability times the loss, we get an expected value for this event.

We can further extend the tree to multiple events involving the next line of defense: Sprinklers. Tree "branching" and probability calculations proceed in the same way. Notice we now have probabilities for annual events ranging from no loss (1 minus.2, or .8), to the catastrophic destruction of the plant. Fault trees are constructed in similar fashion, only they work backwards from outcomes to possible causes.

Obviously the credibility of the data is dependent on the accuracy of the sub-event probabilities and the logical structure of the model.

Ultimately, there is no way to verify the low probability results. As a result, the quantitative output of such logical models, in terms of probabilities, remains very much a matter of faith. The prudent strategy would be to treat such outputs with caution. As a *qualitative* tool for assessing the structure of risky events (consequences only, ignoring probabilities), event and fault trees can still be an extremely valuable risk management tool, even when probability data is very uncertain.

Computer simulation of probability models is sometimes viewed as a way to increase the credibility of statistical results. This view is erroneous, however. Simulations can not increase the credibility of their underlying models. These models can only be specified through induction (i.e., experimentally). Confusion enters in that simulations in the study of risk are often used as a numerical tool to solve complex underlying models. When we can't solve these models analytically, that is, using math, we turn to numerical methods like simulation.

The use of simulation to solve complex equations was developed in the 1940s, where it was used in the emerging field of nuclear physics. It took its place along side a variety of *numerical methods* to solve equations that are too difficult to solve analytically. Numerical solution of equations using simulation was named the Monte Carlo method, for obvious reasons. Not surprisingly, the rise of Monte Carlo methods paralleled that of electronic computing.

Monte Carlo simulations are often used in the statistical study of risk to combine separate probability distributions for the frequency (number of occurrences) and severity (or size) of potential losses. How accurately the simulated result reflects the true underlying mathematics depends on the number of simulation trials. These trials are based on computer generated random numbers. These randomly generated variables are fed into the given equations, and combined. Each trial therefore shows an answer to a segment of the distribution we seek to specify. The larger the number of trials, the greater the breath of "coverage" of the mathematical result we seek, and hence the more accurate our simulation result represents that distribution. While repeated trials more accurately represent the combination of complex distributions, they do not give us any knowledge beyond what the underlying distributions bring to the problem. If the underlying distributions of loss are inaccurate, our simulation result will be an inaccurate representation of the real-world. This despite being an increasingly faithful representation of the *mathematical* results.

So simulate one hundred times, or a million times, the credibility of the result as a representation of the real-world will only be as good as the underlying equations. In the vernacular of computer programming: Garbage in, garbage out.

Simulations provide important guidance in the statistical analysis of risk. They allow us to observe the complex effects of risk management programs, say some sort of retention or deductible on an otherwise fully insured program, and random events whose probability characteristics can be modeled using mathematical probability models. Computerized simulations are also valuable to provide visualization of data – an actual picture of how losses might unfold over time. These provide additional insight, as well as training on the "behavior" of random variables. Simulation also enhances our ability to experiment with the underlying models. It is these benefits that have earned simulation its valuable place in the risk management process, not its ability to magically increase the credibility of statistical results: *Simulating a mathematical hurricane model a thousand times is not the same as observing a thousand years of hurricane experience!* If our underlying equations are based on limited statistical data, no amount of simulation can remedy these data deficiencies.

By applying some of the techniques above, we can sometimes extend the credibility of statistical results. We eventually face the same data limitations with these approaches as we do with sampling, suggesting that while we can extend credibility, we cannot extend it very far. We are always bound by the relevant data at hand, and that might be limited. So, we might apply all these methods and come up with an estimate of the annual probability of a devastating train accident of .004, or one in two hundred and fifty years. Or, based on historical data and geological analysis we might estimate the probability of a serious earthquake as .0022, roughly one in five hundred years. Any such estimates must certainly be viewed with suspicion, as would the results of any expected value cost/benefit analysis based on this number.

1.6 Uncertainty due to knowledge imperfections

Up to this point, we have only considered one type of uncertainty, or "unsureness", of final outcomes: That based on randomness and probability, as manifested in the urn model. Randomness occurs in both the underlying process (due to our inability to specify initial conditions exactly) and the statistical sampling process, where we attempt to determine the structure of the population from limited observations ("samples"). Another, and indeed very distinct, type of uncertainty enters when we consider knowledge imperfection that occur when data is scarce, or dynamic (i.e., ever-changing).

We can define "certainty" to mean that knowledge that pertains to exact predictability of some event or property. We are certain that when we add the integers two and three, the result will be five. The lack of certainty, uncertainty, can arise from a variety of factors. We have already become somewhat familiar with that type of uncertainty we call randomness. This uncertainty is statistical in nature and arises from the inherent variability of events in which some element of control in the outcome is missing. It arises from our inability to completely specify the conditions under which some event occurs. Like when we draw from a well-mixed bowl of colored balls, blind-folded. However, uncertainty can also arise due to limitations in our knowledge that go beyond initial conditions in random experiments. All measurements – including statistics – are subject to this type of uncertainty. Under knowledge imperfection, uncertainty is defined not by probability, but by *possibility*. The possibilities in turn are defined by how much our available knowledge allows us to narrow them.

Probability and possibility can coexist. Consider, for example, two random processes: A coin toss and the "toss" of an ordinary thumbtack. The outcome of the coin toss is either "heads" or "tails" (one distinct side of the coin, or another). From what we have experienced in our lives with actual coin tosses, what we have read in our statistics books, what we have heard from others, and from logical arguments about the symmetry of coins (i.e., flat circular metal objects), we know the probability of a coin landing "heads" is $^1/_2$, or .5. While we can't predict whether the coin will land heads or tails on the next toss, we know that over repeated tosses, the times the coins lands heads will average to .5.

Now consider a less "known" process: The toss of an ordinary thumb tack. We know that the tack can either land on its round head, point up. Or, it can land oblique with its point "down" (i.e., "on its side"). We know that the probability of the tack landing point-up is somewhere between 0 and 1 (i.e., it doesn't always land one way or the other). However, arguments from symmetry are not much help in narrowing things down here, as an ordinary thumbtack is not symmetrical – at least in the way a coin is. We might have some limited experience with finding tacks lying around, but hardly enough to define a precise probability in the way of a coin. The best we can do is specify some rough interval of possibility about the probability a tack will land point-up.

It might be objected at this point that a little experimentation with the tack would reveal its statistical properties with more precision. We could then simply specify all related uncertainty using probability and statistics. The fact remains that, baring such controlled experimentation, our knowledge of tack toss probabilities remains vague. One need not strain to make real-world analogies (e.g., choosing a good restaurant). No practical issues with tack toss uncertainty? Consider a dropped tack on the bedroom floor at night.

In terms of our urn model, knowledge imperfection would be like drawing from an urn of constantly changing composition. Perhaps we are drawing from a bowl with a hole in it just big enough that a few balls may occasionally drop out. Or some prankster may be tossing in a few extra balls when we are not looking. All these complicating possibilities are accounted for when we make the ball draw under carefully controlled conditions. In fact, we need to carefully specify the conditions of our random experiment so that we can be relatively confident in the results (e.g., "the probability of drawing a black ball from the urn is .1"). The real-world offers no such controls. This means probability estimation in the real-world may itself be imperfect, over and beyond the natural element of randomness. Imagine all the possible things that can be used to intentionally affect the outcome of the urn drawing experiment. In the real-world, we face many more possibilities, especially when the processes we are dealing with are complex. It is impossible to account for all of them, thereby assuring that the only uncertainty we will face will be due purely to randomness. Uncertainty due to knowledge imperfection will be there also.

Obviously, the existence of uncertainty due to knowledge imperfections must be taken into consideration in the process of expected value decision-making. We must specify not only an average loss, or even just its sampling distribution. We also need to consider a range of possibilities. If we do not, the decision process will be faulty. Uncertainty due to knowledge imperfection matters. Say that you must counsel a friend on a serious medical procedure. Two procedures are available. Procedure A has been used thousands of times and has a well-documented, average success rate of 90 percent (probability of success = .9). Procedure B on the other hand is somewhat new. Based on the doctor's experience with this new procedure and some limited clinical trials, the best estimate of success probability she can give us for B is that it is "around 90 percent". If pressed to give a single numerical number for the success rate for these procedures, it would have to be 90 percent for both. Using this information alone, we would have to say the success probability for both is the same, and we should be indifferent about which procedure is selected. However, the uncertainty (imperfect knowledge) surrounding procedure B would almost certainly cause you to recommend procedure A to your friend.

Or consider the estimate of probability of a winner Sunday afternoon of two soccer games. In one case, the teams have met many times before, with the number of wins for each being equal. Both have similar talented players and coaches, and are generally considered "well matched". The other game is between two unknown teams, equally inept by all observations. If we had to suggest a probability for a "win" by either team, it may be "50/50", or .5 each. While the probability estimates are the same, the second match contains much more uncertainty. Which one would you rather attend?

Decision-making with uncertainty due to knowledge perfection is very complex. We can not just take just take some middle or average value as typical, lest we loose valuable information about the extent of this uncertainty. Often we must take the entire interval into consideration, with particular attention paid to its extremes (i.e., endpoints). This complicates the simple application of expected value cost benefit based on probability estimates.

1.7 Generalized uncertainty and the "10 percent rule"

To properly use expected value cost/benefit as a tool for risk management, we need to recognize its basic underlying assumptions. Among those, the ability to average over time, and the need to characterize underlying probabilities with some degree of precision. In the real-world, there are limitations on both. These limitations start to manifest themselves when losses get sufficiently infrequent so that the annual probability of loss becomes low.

In attempting to determine the probability "inputs" to the expected value equation, we are sometimes faced with a small database for statistical estimation. As a result, multiple "possibilities" enter. Not only does the data become scarcer, it is harder to identify any subtle dynamic processes that may be at work over time. As losses get more and more infrequent, identifying probability estimates gets harder and harder. We need to consider this generalized uncertainty (randomness *plus* knowledge imperfection) when making decisions about risky events.

Unlike the case with sampling error in well-defined situations, there are no mathematical formulas that express the degree of uncertainty we face when making decisions based on real-world loss data subject to generalized uncertainties. Based on observation of real-world decisions, and a little logical thinking about the practical data limitations we face, we might in fact suggest a rough dividing line past which statistical predictions in ordinary (i.e., uncontrolled) loss situations must be treated with caution.

In the assessment of accidental losses that manifest themselves through time, we can make this specification in terms of the number of years we can go out before severe uncertainties enter. A reasonably "observable" time span in this regard is probably no more than ten to twenty five years or so, suggesting a probability, on the high end, of 1/10, .1, or "10%". As a very conservative rule, decision based on formal risk assessments that suggest probabilities less than .1, or "10%", should begin to garner suspicion: Does data scarcity limit what we can accurately say about these probabilities?

The rough interface between statistical and catastrophic risk, in terms so of annual probabilities of occurrence, are shown in Figure 1.4. We define statistical losses here not just by their frequency, or probability. By association, these losses are also smaller in size (though their impacts certainly cannot be ignored). Catastrophic hazard losses exhibit annual probabilities far lower than .1, or even .01 ("one in a hundred"). Reasonably, they may have probabilities that we can only very imperfectly assess as being as low as one in a thousand, and often much less. This discussion suggests that while tremendously valuable to the overall process of managing risks, the statistical methods that underlie expected value cost benefit analysis may have limitations that must be considered when dealing with high-stakes risks. Frequency and observability define the world of statistical risks. On the other hand, high-stakes risks are characterized by their rarity, and, above all, their potentially disastrous impacts. These conditions severely impact the ability of expected value cost/benefit analysis to provide adequate guidance in the management of high-stakes risks. To understand these risks, and how to deal with them, we need to go beyond statistical methods.

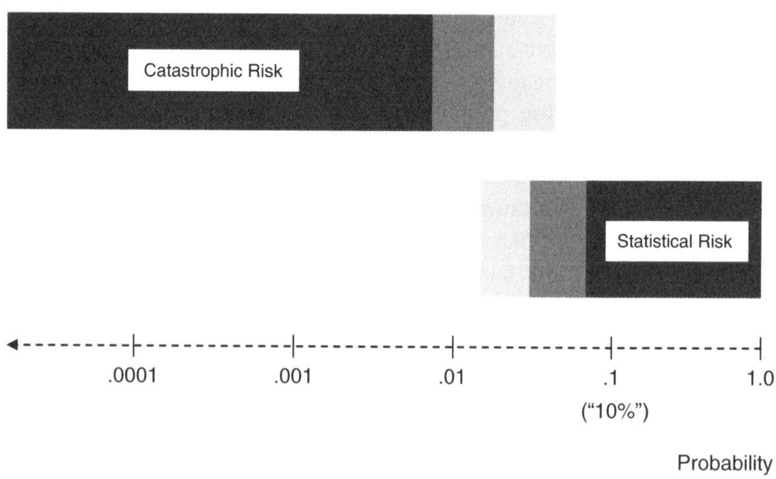

Figure 1.4 The Boundaries of Statistical Risk Treatments

2
The ABC's of High-Stakes Decisions

Attention to the immediate can obscure rare, but significant events, that can have absolutely dire consequences should they occur. The statistical approach supposes that the risks we face present themselves within a relatively short time horizon. At least short enough that the statistical calculations, based on previous history, make sense. In fact, most of the significant risks we face have a fairly low probability of occurrence, and hence, may not be amenable to analysis using statistical methods. This means we need to use specialized decision criteria for dealing with high-stakes risk.

2.1 The "iceberg" model of risk

The totality of risks we face – as individuals, business entities, communities, or society as a whole – may be visualized as an iceberg floating in the ocean. This "iceberg" model of risk (Figure 2.1) suggests that we are most aware of those risks that we can experience within a relatively short-run of time. It is therefore the visible "tip" of the iceberg that commands most risk management attention. From the standpoint of the individual, things that could be treated statistically include catching a cold during the winter season, a flooded basement, an automobile "fender bender", the need for a new set of tires, and the like. From the business perspective, the statistical tip of the iceberg includes worker mishaps in the workplace, product warranty losses, fleet auto accidents, "slips and falls" in public areas, as well as fires, water damage and thefts of property of a "non-catastrophic" level (i.e., those of relatively low financial impact). On a societal level, statistical risks include flu outbreaks (genuine "epidemics" excluded), homicides (sadly), minor floods,

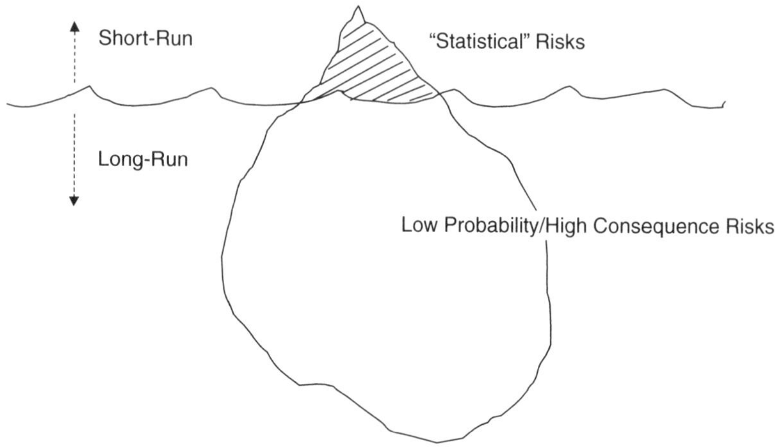

Figure 2.1 The "Iceberg" Model of Risk

windstorms, and other natural phenomenon (again, of a non-catastrophic nature from the perspective of the collective).

It is what "lurks below" that can truly cause the most damage to us, our enterprise, our society and our world. Due to the complexity and dynamics of real-world processes, be they physical, technological, social or political, we cannot "predict" the behavior of these risks in a statistical fashion, and hence they are difficult, if not impossible, to manage. For the individual, these include loss of life or serious disability due to disease or accidents, and serious financial difficulties or bankruptcy, due to a loss of a job, or due to physical or mental illness. For the business enterprise, catastrophic physical events such as fires, floods or windstorms, legal liability imposed for negligence damages arising from products or operations, and financial ruin due to dishonesty or malfeasance of management. Worldwide risks of note include natural calamities, war, and epidemics.

All of these have low probabilities of occurrence, yet their consequences can be devastating. As we will see, the approach to high-stakes decisions, those involving catastrophic events, is much different from that we apply to the more mundane aspects of life. As opposed to being cost-avoiding, high-stakes decisions are often existential. They may not grab our immediate attention, by virtue of the fact that their occurrence is a rarity.

The approach to high-stakes risk management is not easy. With big loss potentials come big issues with respect to how we handle those

Action	State of the World: No "Loss Event"	State of the World: "Loss Event" Occurs
Do Nothing	0	\$X
Take Preventive Action	\$Y	\$Y

Figure 2.2 A Simple Risk Management Decision Matrix

potentials. As a result, the responses to high loss potentials are often more psychological than rational, and often based more on myth, custom and culture than on reality. We won't pretend to come up with any final answers in this primer on high-stakes decisions. Rather the aim is here to open the eyes to some lesser-appreciated, or perhaps unduly neglected, modes of thought with respect to high-stakes decisions.

To simplify our discussion of high-stakes risk, we will limit our risk management alternatives to two: "take action" or "do not take action". These are represented using a simple matrix format, known as a "decision matrix", or "table", as shown in Figure 2.2. While this representation is simple, it does capture the essence of high-stakes decisions. The ideas behind it can be easily translated to more complex representations, for example, when magnitudes of exposure are continuous.

To represent the possible "state of the world" we assume a "loss event" (fire, windstorm, auto accident, ecological accident) either occurs, or it does not. If a loss event occurs, we lose some monetary amount \$X. Our direct actions are simply to do nothing, or take some preventive action (which may include avoidance) at cost \$Y. We assume further that the loss event occurs within an annual time period with probability p_x.

In this simplified example of risk management decision-making, optimization proceeds on the premise that we can balance loss prevention costs against the long-term average, or expected value, of loss (as defined in Chapter 1). That is, a suitable balance point between prevention and probabilistic loss exists when,

$$\$Y = p_x \bullet \$X \qquad \text{(Equation 2.1)}$$

We take preventive action (and accrue a positive net gain) as long as the left side of the equation (cost) does not exceed the right (benefit, in terms of probabilistic loss reduction), where the right side is simply the expected value of loss (Equation 1.4). Following this rule, we assure ourselves that *in the long-run*, the benefit obtained from avoiding losses will exceed the cost to do so.

The application of expected value criteria via the balancing equation is a very straightforward and simple case of cost/benefit optimization. The whole process sounds kind of easy. This apparent ease of application should be our first "tip off" that something is amiss between theoretical application of expected value and real-world application. In the real-world, decision-making regarding high-stakes decision is very, VERY hard. Simple comparisons, such as those suggested in Equation 2.1, do not offer what seem like the "right" answers, at least on the basis of intuition.

2.2 Why expected value decision-making doesn't work

As we have suggested above, when our time horizon is short, the management of risk via expected value, or average loss, optimization becomes a rational and quite practical endeavor. If I recognize that a large department store has incurred $10,000 of legal liability losses over the last five years, because of speeding in the parking area, it will make sense to install speed bumps costing an amortized $1,000 a year to install and maintain. We may reason with relative confidence that losses "average" about $2,000 per year. Our net gain is $1,000 a year. Chances are that we will consistently recognize this gain over a relatively short time horizon (2 to 10 years), thereby verifying that the installation of speed bumps was in fact the minimum expected cost decision. We can manage relatively frequent probabilistic events statistically, based on the notion of averaging losses over time.

Now consider the business that depends on a $1,000,000 store, and its contents, for sales. First of all, to apply expected value decision-making, we need a precise estimate of the probability of loss. As discussed above, when dealing with the low probabilities that characterize high-stakes risk, it is simply not possible to obtain such precision. This is due to a complex and dynamic environment that virtually assures uncertainty due to knowledge imperfection. Making decisions based on the assumption that our estimate is correct, or close enough, can be very misleading. Some precise number may in fact serve as our "best guess", but without any way to specify the associated uncertainty, one guess is as good (or, as bad) as another. Expected value decision-making is simply not geared to uncertainty that arises from knowledge imperfections.

Now let's say that we (somehow) determine the annual probability of a (total) loss to be one in two-thousand, or .002. The expected value of loss is .002 x $1,000,000, or $2,000. Assume further that it costs $5,000 (annually) to install sprinkler protection that would prevent the loss. As the cost of protection is greater than the expect value, the minimum expected cost solution is to forego protection. Yet, if the plant burns down tomorrow, the $3,000 saving is of cold comfort. The firm is ruined. In light of the potentialities, the expected value decision does not seem conservative enough.

When losses are large, catastrophic, and quite possibly ruinous, the ability of Equation 2.1 to "balance" costs and probabilistic benefits is suspect. The simple *potential* for losses is just too large. Attempting to commensurate the two via the averaging process does not work. We are assuming an underlying time-horizon that may cease to exist if the loss occurs.

Our confidence that we can in fact recognize significant risks, and take proper action to eliminate them, is bolstered by what is ultimately a very shortsighted view of the potential risks we face. We understand what we can experience. This results in a very localized view of risk. The more global view is obscured. When it comes to the risk as it unfolds in the real-world, the horizon we can properly experience can be roughly measured in terms of a time. Applying the "10% rule" (Section 1.7), what we can credibly experience in terms of most risks we face usually involves a time horizon that at its maximum is maybe 10 years to 25 years. This translates to a probability of 1/10 (.1) to 1/25 (.04) or so.

The nature of risk therefore relegates statistical management of this risk (reduction or elimination) to a relatively short, local time horizon. Serious risks, however, do not *necessarily* happen within a relatively visible horizon. Rather we may estimate them to occur once in a hundred years, or a thousand years, or even a million. How, outside of the limited realm of controlled experimentation, could we possibly verify such small numbers? Or more pertinent to the true management of risk, how do we know we have genuinely affected (reduced) these probabilities *via* our actions? The best we can do is to extrapolate from short-term experience. The farther into the future we do so, the less reliable such extrapolations.

In understanding high-stakes decision-making we therefore face two issues, both regarding the probability of loss and our ability to assess it. One is the fact that we can only estimate probability of high-stakes loss very imperfectly. In short, the relevant probabilities may be unknown,

The "Catastrophe Problem":

Catastrophic losses are terminal, ruinous, irreversible. There is no opportunity for recoupment or "averaging" of losses through time, because time, or at least our place within it, ceases to exist.

...in the long run, there is no long run!

Figure 2.3 The Catastrophe Problem

or at least very imperfectly known. Second, is what we may call the "catastrophe problem" (Figure 2.3).

When losses are ruinous, the possibility of recoupment via any kind of reasonable "averaging" process does not make sense. In the long-run, there may be no long-run. In this case, probabilities become irrelevant to the decision process.

2.3 Decisions when probabilities are unknown, or irrelevant

Given the difficulties in dealing with the probabilities of high-stakes risks, we may want to consider decision criteria that dispense with the probability dimension altogether. Let's examine what is perhaps the most the most "non-conservative", or permissive, non-probabilistic criteria with respect to risk first: What we will call the *fatalistic* approach. We'll expand on the notion of conservatism in risk decision as we demonstrate the various decision criteria available.

Under the fatalistic approach, we ignore the probability dimension of risk completely, and focus exclusively on outcomes. The fatalistic approach is supremely non-conservative in that it makes no special considerations for catastrophic loss potentials: Whatever happens, happens. The fatalist resigns him or herself to the occurrence, or non-occurrence of loss, and their motto becomes, "Why worry?".

A decision matrix for the fatalistic approach is shown in Figure 2.4. Here, the cost of a loss, should it occur, is \$1,000,000. The cost to prevent the loss (i.e., reduce its likelihood to "0"), is \$5,000. In terms of our decision matrix, we proceed to identify the minimum possible cost associated with the protect/do not protect decision by systematically reviewing our options. If we do not protect (row 1), and there is no loss, we loose nothing. The cost is "0". If there is a loss, we lose (in our hypothetical setup), \$1,000,000. We'll assume throughout our discussion of decision criteria that, for the entity under study, this \$1,000,000 loss is

The Fatalistic Approach

Actions	No Loss	Loss Event Occurs	Minimum Cost	Choice
Do Nothing	$0	$1,000,000	$0	X
Take Preventive Action	$5,000	$5,000	$5,000	
		Minimum =	$0	

The Precautionary Approach

Actions	No Loss	Loss Event Occurs	Maximum Cost	Choice
Do Nothing	$0	$1,000,000	$1,000,000	
Take Preventive Action	$5,000	$5,000	$5,000	X
		Minimum =	$5,000	

Hurwicz
($\alpha = 1$)

Actions	No Loss	Loss Event Occurs	Weighted	Choice
Do Nothing	$0	$1,000,000	$0	X
Take Preventive Action	$5,000	$5,000	$5,000	
		Minimum =	$0	

Expected Value
(Probability of loss = .002)

Actions	No Loss	Loss Event Occurs	Expected Value	Choice
Do Nothing	$0	$1,000,000	$2,000	X
Take Preventive Action	$5,000	$5,000	$5,000	
		Minimum =	$2,000	

Interval Expected Value
(Probability interval = .005)

Actions	No Loss	Loss Event Occurs	Expected Value	Choice
Do Nothing	$0	$1,000,000	$7,000	
Take Preventive Action	$5,000	$5,000	$5,000	X
		Minimum =	$2,000	

	Fatalistic	Precaution	Hurwicz	Expected Value (EV)	Interval EV
Do Nothing	X		X	X	
Protect		X			X

Figure 2.4 A Comparison of High-Stakes Decision Criteria

clearly "ruinous". On the other hand, if we protect by investing $5,000 (row 2), the $5,000 investment is our total cost, regardless of the state of the world (loss/no loss). We take the minimum of each row, finding "0" for not protecting, and $5,000 for protecting. Now we take the *smallest* (minimum) of these as our choice. We choose to NOT protect (i.e., do nothing).

At the other extreme of conservatism is what we might call the *precautionary* approach. Again, we ignore probabilities and focus strictly on the consequences. Now, however, we seek to avoid the "worst case" scenario. In our decision example (Figure 2.4), we would focus on the $1,000,000 loss when examining the "do nothing" option. Under the prevention option, the maximum cost is $5,000. Minimizing the maximum cost of each action, we now choose to protect. This decision criterion is also known as the *minimax* approach.

These criteria are the first things we learn about in any good introduction to decision theory. Cases where the likelihood of loss is either unknown (or irrelevant) are referred to as "decision-making under uncertainty". Most texts on decision theory introduce decisions under uncertainty early on. Contrasting approaches to uncertainty include some variant of the fatalistic approach, and the conservatism of minimax is reviewed. As quickly as they are introduced, however, they are usually dismissed as having little "practical" value. The assumption is that we always have *some* knowledge of probabilities. It is also assumed that the probability dimension of loss is always relevant, regardless of the impact of the loss. There are also suggestions that precautionary avoidance, on the basis of minimax, is simply too "conservative" an approach for application on any practical level. Rarely are the full implications of these offhand dismissals completely examined. A sampling of most texts on "decision theory" show the bulk of the discussion within the text focuses on *statistical* losses.

Likewise, traditional risk management texts usually list "risk avoidance" (what we are calling precautionary avoidance, based on minimax) and "risk assumption" (fatalism) among alternative risk management techniques. Just when and how these techniques are used (and their implications) is given little or no treatment, with the emphasis instead on statistical loss prevention, and insurance.

It is unfortunate that these challenging approaches to risk are so easily dismissed, in favor of the more tractable statistical approach. Given the difficulties we face with applying expected values to extreme decisions, it makes sense to take a closer look at these decision-making techniques, and their alleged defects. Honest and effective responses to high-stakes risks require detailed scrutiny and analysis of these very basic, yet difficult to apply, decision criteria, rather than automatic reliance on easy, yet completely inapplicable, "statistical" criteria.

2.4 The dilemma of precaution

The great benefit of the precautionary approach based on minimax is that for all intents and purposes it eliminates risk. In fact, only the "all or nothing" approach inherent in minimax can assure us protection from catastrophe on a consistent basis. All other approaches admit at least the possibility of devastating losses, and possible ruin. However, the conservatism gained in applying the precautionary approach comes at a cost. Depending on the circumstance, this cost may be expensive. In the example decision matrix shown in Figure 2.4, we may view a $5,000 expenditure as being fairly modest with respect to preventing a $1,000,000 loss (probability considerations aside). Realize, however, that precaution, based on minimax, would suggest that we invest in protection even if the cost were $999,999. Under precaution, we must be prepared to spend up to the amount of loss to prevent a loss. The result could, in and of itself, be ruinous. This presents us with what we may call the *dilemma of precaution*.

The paradoxical nature of precaution is evident in its use in the management of what is often suggested as one of our most pressing global risks: The potential for global warming. Considerable scientific evidence points to the fact that the mean temperature of the earth is rising. The exact cause of this warming, or even substantive evidence of warming as a genuine trend, as opposed to a cycle, has yet to be established with a high degree of confidence. If such a prognosis is true, however, the results on humanity and the environment would be truly catastrophic, as temperature holds a delicate balance for life on earth. Precaution demands the curtailment and subsequent elimination of activities that may cause such warming. One very plausible cause is the increase in so-called greenhouse gases in the atmosphere. These gases, mostly carbon based, such as carbon monoxide (CO) and carbon dioxide (CO_2), are a byproduct of industrial production. A cutback in these gases is therefore linked to a cutback in production. While a setback to the progress of industrialized nations, such curtailment could mean a complete halt to "modernization" of less developed nations. The opportunity costs of greenhouse gas cutbacks are obviously tremendous. And this is just the direct costs. Might not the curtailment of greenhouse gases, and hence development in the already underdeveloped countries lead to political unrest, and perhaps pose an equivalent threat to world safety due to war or terrorism? Such dilemmas often lead people to pronounce that precaution, while well meaning, is simply unworkable as a bona fide risk management alternative.

2.5 Can precaution make things worse?

It is sometimes suggested that precaution based on minimax is not only expensive, it can actually make things *worse,* by ignoring what are known as "risk-risk tradeoffs". By avoiding one activity or action that creates risk, we may be promoting yet another risk, or risks. Consider the case of what amounted to a precautionary ban on DDT in the 1980s. While the case for DDT as both a human carcinogen and a threat to the natural environment was not (and is not) conclusive, there was sufficient information to suggest that there was at least a possibility of catastrophic impacts from continued, widespread use. On the other hand, DDT is a particularly effective pesticide against mosquitoes, a common carrier of the malaria virus. Might the precautionary ban on DDT make the overall risk greater by removing from the market an effective preventive measure against the spread of malaria and other insect borne diseases? Perhaps. This is something that must be investigated as part of the overall precautionary risk analysis.

Nothing in the precautionary approach suggests we abandon science. We just have to respect its limitations. If the disease threat becomes the greater of the two, the minimax becomes "use DDT for disease prevention" (as opposed to "ban DDT"). When the decision model is correctly specified, minimax decisions can not make matters worse, *because the worst-case becomes the focal point of the decision.*

In his paper "Endangered Species and Uncertainty" (*American Journal of Agricultural Economics*, 1978), economist Richard C. Bishop examines a precautionary approach to the management of ecological risks. He begins his analysis by assuming that a hydroelectric dam project is proposed which would flood the last remaining habitat of an endangered species. Bishop then sets up a decision matrix that looks very much like the simple matrix we presented in Figure 2.2. Dam construction presents the possibility of a catastrophic loss of species diversity, to which we assign a value \$X. Preventing species extinction means instituting what Bishop calls the *safe minimum standard* of conservation, or SMS, and requires that the dam project be halted. Doing so would result in "opportunity costs" in the form of the lost benefits the hydroelectric project would bring. We show these costs as \$Z. To a developing community, these costs could themselves be quite substantial. From the opportunity cost perspective, if the project proves harmful, the possible loss \$X must be adjusted by the net gain of species protection (i.e., avoidance of catastrophic loss of \$Z).

		State of the World	
		No "Loss Event"	"Loss Event" Occurs
Action	Do Nothing (Allow the Dam)	0	$X
	Take Preventive Action (SMS)	$Z	$Z–$X

Figure 2.5 Extending the Decision to Consider "Opportunity Cost"

A modified decision matrix, including opportunity costs, is shown in Figure 2.5. Applying the minimax criterion we find that if the opportunity costs, $Z, exceeds the cost of harmful effects ($X), we proceed with the hydroelectric project. On the other hand, if $X exceeds $Z, we *do not* proceed with the project (enforcing the safe minimum standard). Obviously, the decision is sensitive to how we value the respective outcomes. We will return to issues of valuation later. Application of the precautionary criterion, however, remains straightforward.

The mandate that we use full available, credible scientific information on costs and benefits in making precautionary decisions is no different than in statistical decisions. Precaution does not ignore benefits – it ignores probabilities in the net balancing of cost and benefit (i.e., expected values). It does so because they provide no useable informational value. The complete range of costs and benefits can be easily accommodated, as can the notion of risk-risk tradeoffs (e.g., DDT vs. malaria, hydroelectric power vs. species extinction).

Precautionary dilemmas only enter the picture when the opportunity costs of precaution are very high. In other words, minimax dictates precaution even when significant benefits are foregone to protect against catastrophic loss. The principle *does not* suggest that forgone benefits should ever be allowed to *exceed* potential loss costs.

Of course there may be cases where the consequences of precautionary actions are themselves unknown, leading to genuinely disastrous results. We may take precautionary action and later be completely surprised to find the actions have failed, or even worsened the situation, because the decision alternatives were imperfectly framed. Once again, we face potential pitfalls that are not unique to precautionary decisions. Statistical decisions may also fail to consider the full range of options, due to ignorance factors, with unintended, and possibly quite bad, results. There is nothing inherent in the precautionary approach that makes such ignorance more prominent than in statistical decisions.

While cases like these show that precaution cannot make things worse, they clearly show once again that precaution can be very costly, both in terms of direct cost and opportunity losses. That is, it may not make things better either. When the costs of both prevention and cure approach enormous proportions, we face the dilemma of precaution. We are in effect "doomed if we do, doomed if we don't". The decision between the two becomes arbitrary: A coin flip. We could go through the expense of a full-blown statistical analysis, but the data will not be credible. So what have we gained? Judging that one or the other catastrophe is "more probable" is fraught with uncertainty. Even if we could determine the probabilities, how would we justify picking the less likely one? Only *via* statistical averaging, which is, as we have pointed out repeatedly, inapplicable in the case of catastrophe (...there is no "long-run"). Any rewards we receive from choosing the less likely would therefore be strictly "psychological". The dilemma exists whether we like it or not. It's not the "fault" of precaution. It is the way of the world. To attempt to make these dilemmas go away by introducing a more liberal approach to catastrophe, via expected value cost/benefit, or whatever, does not make logical sense. The methods don't stand up to logical scrutiny. Are they worth the value of fooling ourselves into believing they do?

2.6 Modifying expected values for imperfect knowledge

Recognizing the potentially extreme nature of precaution, there have been attempts in the study of decisions to "balance" the conservatism of precaution with the fatalistic approach. One of these is the Hurwicz criteria, named after its inventor, Leonid Hurwicz. Hurwicz himself was one of the first modern decision theorists to recognize the deep issues posed by decisions under extreme uncertainty.

It is interesting to note that Hurwicz viewed what we are calling the fatalistic approach as an "optimistic" criterion, and precaution as "pessimistic". The idea that fatalists are "optimistic" would make sense if catastrophes never happened. If that was the case, it might make sense to hold out the hope that they never will. Unfortunately, catastrophic events have happened through history. This would suggest that the view that they will somehow cease to happen in the future would be a very cockeyed form of optimism, at best.

Under Hurwicz's criteria, an "index of optimism", α, is selected between 0 and 1. The decision outcomes are weighted using this index. An index of 1 gives the same result as the fatalistic approach (absolute cost minimization), as shown in the illustrative example in Figure 2.4, above, and an index of 0 gives the same result as precaution. Indices in between act to blend the approaches.

A drawback of the Hurwicz approach is that the choice of index value is completely subjective. While clearly an attempt to temper the conservatism of the precautionary approach, the Hurwicz criteria has no theoretical basis on which to recommend α, leaving us, essentially, nowhere. The Hurwicz index α, when computed "backwards" from the suggested actions of the decision-making criteria, may have value as providing an indicator of just where along the spectrum from fatalism to precaution a decision falls. In this way, it provides an index of the conservatism of the decision.

We can also introduce conservatism by taking into account knowledge imperfections, via the use of intervals of uncertainty. In this way, we maintain the benefits of the expected value approach, while accounting for uncertainty due to knowledge imperfections. In the example of our individual plant, perhaps we more realistically represent the probability of loss as a range of say, .005 either way of our "best guess" estimate (.002), as shown in Figure 2.4.

In Figure 2.6, we show a simple interval of uncertainty along the probability scale. When we get into the practical analysis of catastrophic risk, we are working with some very small numbers, like "one chance in a thousand", 1/1,000 (.001), or even "one chance in a million", 1/1,000,000 (.000001), or lower. To avoid all the zeros it is convenient to express these numbers using a logarithmic notation, where a negative subscript denotes inverse. So, 1/100, or .01, would be written as 10^{-2} (i.e., $1/10^2$). In this way, even very small numbers can be written conveniently. For example, one chance in 8 million (1/8,000,000) is written as 8×10^{-6}.

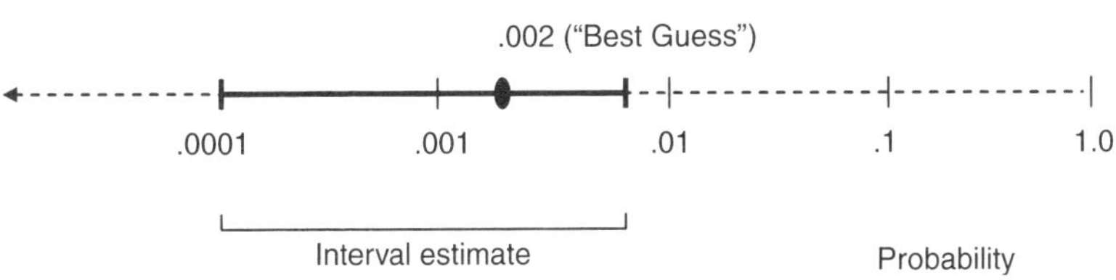

Figure 2.6 An Interval Estimate of Probability (Including "Best Guess")

The scale in Figure 2.6 is marked in logarithms of $1/10^{-x}$, so that we can fit some very small probabilities all on the same scale. While .002 is our best guess, the arrows indicate an interval of possibility that extends towards both sides of .002, between .0001 and .007. All probabilities in between indicate genuine "possibilities" for the true probability value.

Note that though similar in appearance, possibility intervals are far different from statistical confidence intervals. What we attempt to capture here is uncertainty due to knowledge imperfection, not randomness in the statistical sampling process. The latter is a result of drawing samples that constitute some subset of the entire population of results. The smaller the sample, the less representative it is of the wider population. Calculating statistical confidence intervals amounts to assessing the possible combinations of results that can occur on random drawings of sample size x, from a wider population of size x + y. The uncertainty that results is all a function of randomness, i.e., drawing blind from a "mixed" urn. In addition, the common interpretation of such intervals as a "probability of a probability" becomes problematic in the face of uncertainty that is not properly treated as a form of randomness. Uncertainty due to knowledge imperfection is different from randomness, and must be treated differently.

Back to our example: If we take the extreme endpoint of our interval estimate (.007) as our conservative guess of the probability, the expected value of loss calculates to $7,000. Protection now becomes the best option. Note however, that intervals do not guarantee what we might intuit as conservative action with regard to risk. If our interval had been pegged at ±.002, rather than ±.005, we would have continued to reject the loss prevention action based on the interval valued expected value, with the resulting discomfort as to the conservatism of our choice.

Recognizing the "epistemic" uncertainty in probability estimates results in naturally more conservative action with regard to risk. For example, extending the boundaries of our expected value risk estimates to take into account uncertainty will make more protective measures economically feasible within the framework of expected cost versus benefit comparisons. But exactly how much more conservatism does the recognition of uncertainty provide? Or, more to the point, are the revised measures conservative *enough*? There is nothing in the logic of interval estimation itself that can answer this question.

Perhaps there is something in the individual psychology of decision-makers that can give us some idea of the "proper" degree of conservatism when making high-stakes decisions. Let's examine what decision-makers reactions to risk might suggest about the way they approach monetary losses, or perhaps the assessment of relative probabilities.

2.7 Utility and risk aversion

Experiments and observations of "real-world" decision-makers have suggested that these decision-makers "value" larger losses disproportionately higher than lower ones. This valuation can be expressed in terms of a negative utility, or disutility, measure based on some weighting of the monetary value of loss. That is, we use, say, a dollar metric, but weigh that dollar amount according to an equation that represents the decisions-makers distaste for larger losses.

The idea is shown if Figure 2.7. We show disutility as measured in dollars against a scale of loss (also in dollars). When disutility is valued equal to loss, the decision-makers' approach is what is called risk neutral. That is, they behave strictly on the basis of expected value. On the other hand, if larger losses are valued more, the decision-maker shows what may be termed risk aversion. Clearly, if some

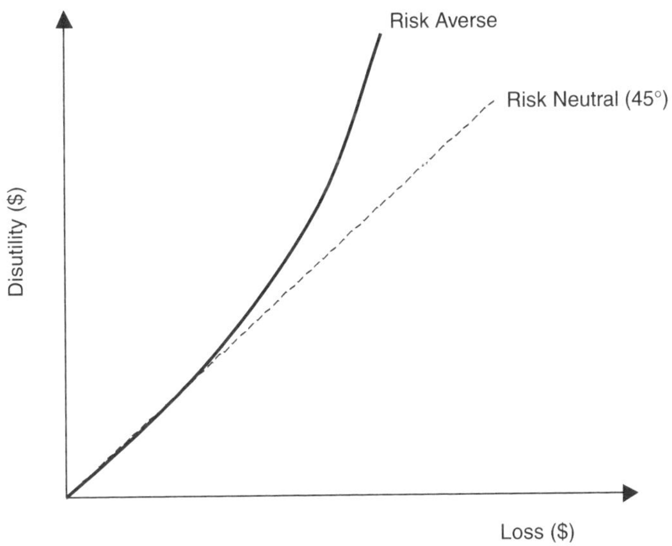

Figure 2.7 Disutility and Risk Aversion

degree of risk aversion is considered, expected disutility of a risk exposure will be greater than expected utility for any given probability of loss. Risk aversion makes the decision-maker more conservative.

The idea of disutility can also be combined with interval probabilities, to take into account knowledge imperfections about probabilities. Instead of an interval based expected value we now have an interval based expected disutility. The math is straightforward.

The problem with disutility measures is that they are very subjective. Disutility curves are usually fitted to observed decisions. Does disutility "explain" the decision, or *vice versa*? That is, if decision is based on disutility, what is disutility based on? While the general idea that decision-makers value larger losses more than smaller ones makes intuitive sense, measurements remain entirely subjective. We have introduced conservatism, but how do we know we are being conservative enough? Subjective arguments in this regard quickly become circular. Our community instituted a protective measure against risk, despite the fact cost exceeded expected value. Their disutility of loss must have been higher. We infer utility only after decision. But why not use such decisions to define the disutility function, as is often done in experimental studies of disutility under controlled conditions? We could then use it to make future decisions. But let's say another community, under identical probability/loss conditions did not undertake protection. Then their disutility is lower. But how do we reconcile inter-subject differences in utility? To use utility as a guide to group decisions under any consistent basis is extremely difficult. And what have we gained, except a descriptive theory of decision. It shows how decisions are made, not why. In the end, we have just substituted a psychological construct for a parameter that has perhaps no further justification than the choice of the α weighting in the simple Hurwicz criteria described above.

Along the lines of modification to the consequences, some risk analysts have suggested weighting to the probability side of the expected value equation. Instead of weighting consequences, in what some might say is a fairly artificial manner, why not weight probabilities? The use of weighted probabilities was first suggested as a way to avoid some anomalies associated with observed decisions based on the construct of weighted consequences (disutility).

Figure 2.8 shows a function that weights probabilities rather than consequences. This "inverse – S" shaped curve provides a greater weighting to smaller probabilities (regardless of size of loss). The degree of weighting is determined by adjusting a weight parameter, λ.

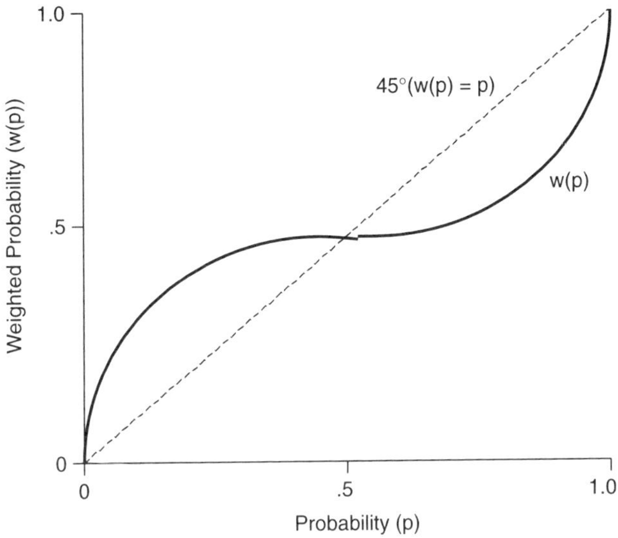

Figure 2.8 A Weighted Probability Function

The weighted probability is used in place of the "measured" probability in the expected value function. For the low probabilities normally associated with catastrophic losses, application of this weighting function has the effect of making risk decisions more conservative. That is, the "expectation value" of loss so modified will be greater than the unmodified value suggesting that "optimal" loss prevention expenditures will be commensurately greater, per Equation 2.1. Once again, we are faced with subjective modifications, with the parameter of the modified probability equation determined solely by the "psychology" of the individual.

Both modification of the consequences or probabilities in the expected value equation still leave us dependent on the idea of averaging. With respect to averaging catastrophic losses, we are faced with the catastrophe problem: In the long-run, there is no long-run. If we choose any decision point less than the amount of catastrophic loss, we leave ourselves open to terminal failure. Expected disutility can only manifest itself over time, just like expected value. We are still dependent on the idea of averaging. And in light of catastrophic potentials, averaging just doesn't make sense.

Risk aversion, in terms of disutility, in and of itself does not provide a sufficient measure of conservatism with regard to high-stakes loss.

Expected value approaches, even when suitably extended to account for uncertainty in the probability estimates, disutility of outcomes or differential valuation of probabilities, may not present an approach that is conservative enough with respect to the possibility of truly momentous losses happening. This brings us back to our most basic criteria, fatalism and precaution.

2.8 Where does insurance fit in?

What about a very commonly used "tool" for risk management: Commercial insurance? Hundreds of millions of dollars change hands each year, between insurers and policyholders, based on random events. It is quite obvious that commercial insurance avoids quite a few of what might otherwise be considered catastrophic failures, at least those of a financial nature.

Commercial insurance and other organized "pooling" mechanisms can expand the collective time horizon of risk taking, by subscribing multiple pool members ("insureds"). Averaging of losses can be stretched further in this collective atmosphere, than it can by any individual "insured". Insurers also sometime provide advice on loss prevention expenditures. Can their broad experience be used to widen the spectrum of loss experience? Perhaps. But while insurers themselves engage in so-called loss prevention and risk management activities, most of these are simply an adjunct of achieving optimum application of insurance rates through effective risk classification. Less protected exposures to risk earn higher rates and hence premiums. While this may in fact be an indication of higher risk, and the physical difference of higher risk versus lower risk exposures may to some extent be identifiable, altering this risk profile may not be a matter of discretion.

From a property insurance standpoint, for example, chemical production plants will always be more risky than dairies by virtue of the fundamental processes involved. And while the property risk within a class of chemical plants may be reduced by effecting a greater degree of physical loss prevention and protection, the exact correlation between prevention efforts and premium savings is often very tenuous, if it can be shown at all. Recourse to expected value arguments in this regard is subject to the pitfalls previously elaborated.

Insurance is ultimately a pass-through mechanism, not a mechanism to promote genuine risk control. Insurance premium is a very imperfect reflection of risk. It captures relatively short-term statistical regularities, and, as suggested above, can be easily overwhelmed by catastrophes. As a result of not being able to properly capture potential catastrophic elements on a wider scale it does not provide a true measure of risk that can be used as a basis for serious loss prevention decisions. Doing so can be very misleading.

Insurance remains a mechanism that treats risk statistically. While it has demonstrated its ability to alleviate the potential for catastrophe at many levels, it can itself be overwhelmed by events of a sufficiently grave nature. This is evidenced by recent natural occurrence, such as windstorms and earthquakes, as well as by the terrible "man made" disaster that befell us on September 11. Less visible, but as damaging to insurer's capacity to manage risk, have been the cumulative effects of pollutants such as asbestos. In a principle akin to the physical law of the conservation of matter, risk cannot be destroyed. It can only be divided up. While the immediate effect of the insurance pooling mechanism is to lighten the financial load of the pool members (insureds) having losses, ultimately it just amounts to the redistribution of these losses.

While insurance will always remain part of the risk manager's toolkit, it is not the final solution to the issues of unknown, high-stakes risks. It takes us, at most, only slightly "below the surface". Reliance on insurance alone cannot in and of itself constitute a proper response with respect to how we handle risks. Be it in a personal, business or societal setting, a comprehensive policy with respect to risk is needed. In managing risk, an insurance policy is never a substitute for a comprehensive risk policy.

2.9 Fatalism, by default?

Expected value criteria do not make sense when applied to catastrophic risk, due to what we have identified as "the catastrophe problem" (Figure 2.3, above). Insurance and other pooling mechanisms extend the reach of expected value methods somewhat, but their domain is limited. Precaution, which ignores the probability dimension of loss altogether and focuses strictly on consequences, has its own potential drawbacks. On the face of it, it would seem that some sort of fatalism with respect to risk might be inevitable.

The true goal of risk management is to reduce our concern and worry over risk. Ostensibly, it does this by analyzing the risks we face and either eliminating them or reducing them to an acceptable level. That said, we have to recognize that there is an alternative for reducing or eliminating the worry over risky events. This strategy is based on a sort of rational, "pragmatic", fatalism regarding risk: We cannot do anything about the truly significant risks of the world, so why worry?

While the connotation of fatalism is almost universally perceived as negative, the facts of real-world risk management make a strong case for at least some sort of conciliation in the face of the unknown (and unknowable). Despite its negative connotation, being realistic about fatalism is certainly preferable to laboring under misguided optimism as to what we can and cannot know and do about risk.

The first step in divining the future of risk management is to set our understanding of risk and its treatment on a more realistic basis. To do so, we have examined some common approaches to risk and its treatment and noted their shortfalls. Understanding and accepting these shortfalls requires a fair bit of openness on the part of the traditional risk manager to evaluate what has come to be the accepted wisdom in the field. We would suggest here that this acceptance has not always been comfortable. Certainly, for those that feel this discomfort, now is the time to be more vocal about it. Real risk management is not about slogans or personal and professional agendas. It is an endeavor that requires us to predict the unknown. Unfortunately, our ability to do so is limited not only by our individual capacities, but by a complex and dynamic world that rarely stands still long enough for us to get a good read on it.

Despite the comfort we gain from the precautionary approach, the conservatism it entails comes at what is potentially a very great cost. So it seems that if such arguments effectively suggest that genuine management of risk when it matters most cannot be made truly effective without incurring costs that may themselves ultimately lead to ruin (in one form or another), it seems like some sort of fatalism does present a rational alternative of sorts. Of course, our pride in our ability to get things done, face tough challenges and never give up in the face of adversity means that we might not be so ready to admit to a fatalistic philosophy of risk. Spending considerable time and effort managing risk with a short-sighted philosophy based on expected values and the like, however, does not seem a very attractive alternative, short of serving as some sort of emollient in the face of our ineffectiveness.

The fact that we do spend money and time on expected value-based risk management for risks that clearly do not fit the criteria for this sort of treatment (i.e., long-term, catastrophic risks) does not mean that we are not ultimately fatalistic in our philosophy of risk. Some modicum of activity based on expected value in this area would seem to make sense if only to show that we are doing something. Some of this activity may be based on regulatory requirement, some purely on accepted custom. In any case, when expected value is used to manage serious risk it is not used based on its merits. So while "why worry" may look like a callous credo with respect to risk, "do the best you can" at least looks better. Both reflect the same underlying philosophy. There will always remain individuals that both make and accept recommendations on the management of significant risk based on expected value and related criteria, who are convinced of its effectiveness. That said, such a position, even if genuinely held, is not compatible with reality. A realistic view of risk demands some belief in the idea of immediate powerlessness in the face of significant risk.

Based on the facts, fatalism may in fact be the dominant response to the truly significant risk in the modern world. The conclusion seems inescapable when one observes the degree to which superficial thinking about risk, or no thinking at all, dominates much of human endeavor. Too much of what we see in the profession of risk management is based on mere sloganeering: Reduce risk! After all, isn't that what a risk manager is supposed to do? Recognition of what we can know and can not know about risk, and, in turn, what we can or can not do about it, is overdue.

For those that may still find fatalism unacceptable, for whatever reason, there may still be hope in some reasonable modifications to the principle of precaution, or the way it is *applied*. We shouldn't give up on precaution too easily. The problems in applying precaution may not lie in the concept of precaution itself, but in the current structure of the world (as we have made it), and our responses to it. In what follows we will examine such modifications, in the hope that precaution may be at least viewed as a more workable alternative to fatalism, and not dismissed out of hand.

3
Practical Precaution

Actually, the way toward making precaution "workable" may lie in some clarifications, and perhaps some simple modifications, to the principle of precaution. For example, the applicability of the principle, and hence the potential for dilemma, is greatly extended if we insist in the absolute certainty of safety in everything we do. A workable approach to precaution suggests we reasonably relax the criteria of perfect, or absolute, certainty in making precautionary decisions. In addition, many have suggested that at least some element of "costs" may need to enter the principle, in terms of a rough proportionality of precautionary spending with regard to benefits. Ultimately, however, a deeper appreciation of precaution may require that our philosophy with regard to risk, and the activities that expose us to risk, change. This in turn requires a greater appreciation of progress and its implications (good and bad). We will explore these ideas here.

3.1 Is *everything* risky?

One key element of precaution must be clarified before any precautionary approach can be made workable, at any level. That is the notion of what does, and what does not, qualify as "possible". To apply precaution in any type of selective, and hence useful, fashion, we need to have criteria that separate the "possible" from the "impossible". Otherwise, *everything* is risky, and the precautionary approach implies *everything* should be avoided. This would of course make precaution a "non starter".

While a common objection to precaution, the "analysis paralysis" suggested results from a misunderstanding of how a wider view of uncertainty affects how we make decisions about risk. In fact, most of us make decisions about high-stakes risk every day with no particular sense of distress.

Consider the activity of drinking water. Drinking sufficient quantities of water is an indisputable condition of life. Yet, drinking too much water can actually have detrimental, and possibly quite serious, health effects. The social peril of "over-drinking" water, however, cannot be considered a serious possibility for widespread catastrophe. Drinking water would therefore never "make it" to the stage of minimax evaluation terms of precaution.

Now if we focus more specifically on the *quality* of our drinking water, and its delivery system, we may indeed face possible precautionary decisions. Drinking water supplied to the "tap" via municipal water systems is not without its perils. The possibility of harmful contamination exists, and actual serious health concerns concerning municipal water systems surface from time to time. Simple precautionary approaches to the personal management of this risk exist, such as drinking purified ("bottled") water.

There are perils that present a distinct possibility of serious harm under the specified circumstances, and others that do not. Everything is *not* a candidate for precautionary treatment. On a world-wide scale, perils such as global warming, terrorism, natural disasters and pandemics are among those that suggest that precautionary risk management may be worthwhile. On the level of the individual business organization, environmental pollution, serious workplace perils, and development of complex technological products and services are areas were precautionary scrutiny may be beneficial.

How do we formally distinguish between possible and impossible events? Clearly, requiring a deterministic solution (probability of "impossible" events = 0) is too strict for practical application. A sensible definition requires at least some element of chance, or probability. As we have suggested above, probability is very difficult to measure in complex environments, especially when it is small. This means that any practical definition would have to treat a probability threshold for possibility as approximate (i.e., "fuzzy"). An imprecise, yet still useful, definition of practical impossibility in terms of probability numbers, is at least intuitively plausible.

The idea of a probability threshold for "practical impossibility" is the foundation of a risk management strategy based on the concept of *de minimis* risk. The idea behind de minimis is that some risks are sufficiently remote that they can be safely ignored. The principle gets its name from a term of equity law, "de minimis non curat lex" – the law does not deal with mere trifles. De minimis assumes some sufficiently low probability of occurrence threshold has been met, below which we are not concerned.

Decision-making based on some threshold of impossibility based on probability begs the question: How do we determine this threshold? First of all, as we have noted, the concept of "impossibility", or "de minimis" risk, is too complex to permit any single, precise threshold. The threshold is, therefore, more properly specified as a range of possible thresholds, in terms of probability numbers. This range of uncertainty suggests possible thresholds that cannot be excluded by the current state of our knowledge. The idea is the same as the intervals we applied to probability measurements above. Only now, we apply it not to a measurement, but to a concept. Though imprecise, this rough threshold can provide valuable guidance in the way we treat potential risks.

To be useful, we at least need to get an approximate idea of where this threshold may lie. One way to attempt to circumscribe "impossibility" on a practical basis is by examining the history of man. That is, certain exposures, such as the chemical compound we commonly call water (H_2O) has been around since the dawn of man, with no ill effects noted through this history. Likewise, a certain "background" level of natural radiation exists on the earth. Mankind has nonetheless survived. This all suggests we set the threshold somewhere around the inverse of the reasonable history of man on earth. This might indicate a span of 100,000 years, more reasonably maybe 1,000,000 years, perhaps even more. These time frames translate to probabilities of 1/100,000 (10^{-5}), and the proverbial "one in a million" (1/1,000,000, or 10^{-6}). If no effects are demonstrated in a million samples, we might reasonably assume they are not there (or at least, we should not worry about them).

It has also been suggested that a reasonable measure of the threshold of "practical impossibility" be defined in terms of the occurrence of rare accidents that may be called acts of God. Fatal lightning strikes are a exceedingly rare occurrence under normal circumstances of life. Getting killed by lightning is not an impossible occurrence however, and in any given year some small number of people in the United States die from direct strikes by lightning. However, there seems to be no reasonable mechanism to prevent some number of fatal strikes, other than the use of some extremely common sense precaution for not inviting the peril (e.g., do not fly a kite in a thunder storm). The individual chance for mortality from lightning strikes is roughly "one in a million".

We should note at this point that probabilistic thresholds for "possibility" suffer from their own logical difficulties. For one thing, they all are subject to the so-called "lottery paradox". Assume we establish a lottery with many participants. We can establish some arbitrary, but very low, probability threshold, below which we would consider winning "impossible". Yet, in a lottery, someone must always win. Therefore, winning at any probability other than strictly zero, *is* possible, and a logical contradiction exists. Following our lightning example above, we may believe that the chance of getting killed by lightning is sufficiently small (one in a million) to be "impossible". Yet some number of people do die from lighting in any given year. It appears that any threshold approach to possibility vs. impossibility requires a degree of fatalism (i.e., "risk acceptance"), at least at some level.

There is also the issue of verification. We have suggested that when probabilities fall below the 10 percent rule, identification and verification become difficult. Possibility reintroduces probability, but at a very general level. It implies a simple "go/no go" type decision: an ordinal assessment, rather than one relying on cardinal measurement of numbers. Is the probability, roughly, less than some vague (yet useful) threshold? Intuitively, it would seem such decisions require less information than precise estimates of probability, and the subsequent application of expected value comparisons. We take our next step with relative confidence the earth will not open up and swallow us, not with some exact probability assessment that this will be so in mind.

Perhaps our judgments of possibility of risk are not based on probability at all. Physical arguments may be logically combined to assess impossibilities based on physical laws. Limits of possibility may be based on complexity arguments instead: Increasingly complex combinations of events are less possible. In assessing physical possibilities, formal structuring of large impact scenarios may proceed on the basis of event trees, which show the logical progression of loss from initiating event to possible outcomes. The interconnections among events in these trees would have some sort of deterministic, or directly possibilistic structure, rather than a probabilistic one. Both catastrophic potentials and their possible mitigation can be treated in this fashion. Clearly, more research is needed into how we effectively discern the "possible" from the "impossible" when it comes to risky events.

3.2 Defining the "precautionary region"

Just as the probability dimension of risk is uncertain, so is what we may define as "catastrophic". Once again, uncertainty due to knowledge imperfections enters. We can rarely, if ever, specify the catastrophic loss level with precision, be it in term so money, human lives or degradation of our ecosystem. Once again, only a rough-dividing region can be identified.

Recognizing the uncertainty inherent in these thresholds of consequence and probability, we can now, roughly, specify the criteria for the application of precaution. To be subject to precaution, an exposure must be catastrophic and possible. In terms of logic, we say that the "precautionary region" of risk is defined by the intersection of the concepts "catastrophic" *AND* "possible", as applied to some potential risk exposure. This intersection, in terms of our "probability/loss space", or *risk map*, is shown in Figure 3.1. More colloquially, we may refer to this region as the "danger zone".

Having defined the precautionary region, the exercise of precaution becomes a two-step process. First, we determine if an exposure falls

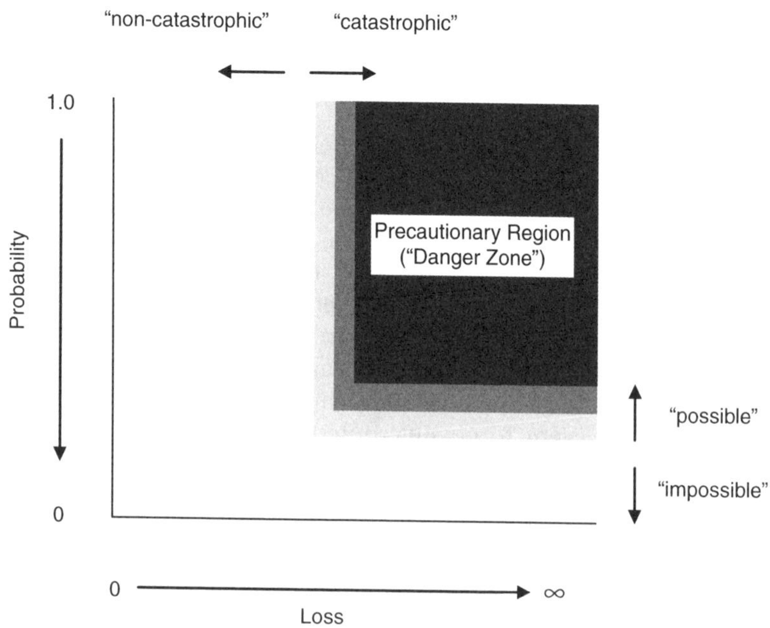

Figure 3.1 Identifying the "Precautionary Region"

into the precautionary region, as defined above. If so, i.e., we determine that "catastrophic" losses are indeed "possible", we then apply the minimax criteria to determine what loss prevention, or avoidance, options make the most sense in terms of long-run survival. In other words, *we avoid the danger zone.*

While the boundaries of the precautionary region are imperfectly known, they are nonetheless useful approximations. The key to effective catastrophic risk recognition, and ultimately, its treatment, lies in being able to assess these approximate dividing lines. We have stressed the importance of further research on defining this region of the probability/loss continuum above. The possibility of identifying this region in rigorous terms lets us apply a variety of analytical techniques to the idea of possible catastrophic risk. Among them, the use of computers to help us model this risk. We will return to the idea of computerized modeling of the precautionary region in Appendix A.

3.3 Integrating measurement uncertainty

In this view of risk recognition we take into account only the "possibility" of harm. The degree to which we have to recognize, or "measure", probability is fairly limited: Is the probability of harm high enough to suggest the possibility of risk? While the probability assessment required is far less stringent than for the effective application of expected value criteria, measurement uncertainties may still have a significant effect on decisions. Not only will our definition of possibility be imprecise, so will our assessment of the threshold probability. If our probability measurements become very uncertain as data get scarce, it makes sense that any scientific analysis of the possibility of harm will be imprecise as well.

Because the assessment of probability of an exposure, as being either possible or impossible, is imprecise, we have to modify the strict application of the precautionary rule. Specifically, if the *range* of uncertainty about probability of harm measurement extends into the region of "possibility", use precaution. Figure 3.2 shows this case in terms of an interval estimate of an uncertain probability.

We only show the probability dimension here, assuming the exposure in question has an unequivocally catastrophic potential should it occur (a large asteroid striking the earth, for example). In terms of probability, we show the precautionary region here as bounded roughly by the "one in a million" (10^{-6}) loss. Higher probabilities fall into the "precautionary region" and require action.

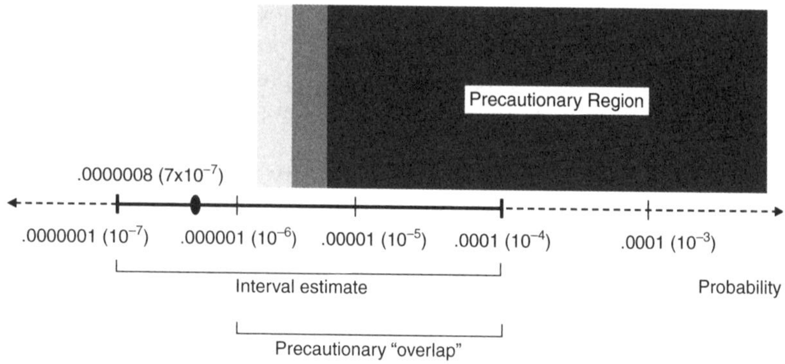

Figure 3.2 The "Possibility of Risk"

Our assessment of the probability of loss is represented by an interval of uncertainty. While our "best guess" of probability lies below the threshold, there is a possibility it can exceed the threshold as well (the precautionary "overlap"). In effect, we have identified a "possibility of risk" requiring precautionary action. In keeping with the conservatism of the precautionary approach, the analysis would suggest that action should be taken when our assessment of the interval probability extends in the precautionary region. It recognizes that we may not be able to measure possibility (in terms of threshold probability) with precision. We must therefore account for this uncertainty about possibility of catastrophic exposures.

While taking into consideration measurement uncertainty once again extends the field of potential precautionary candidates, we would argue, based on practical observations, that there remain actions and activities that we can clearly distinguish "impossible". Even recognizing the effects of knowledge imperfection on precaution, everything is *still* not risky.

3.4 Taking "reasonable" precautions

While a reasoned approach to precaution suggests that "everything" is not risk, a lot of things still are. The dilemma of precaution presents itself in the fact that, on the face of it, widespread application of precaution may simply be too expensive. By acting with precaution, we have eliminated one risk, yet in effect substituted another – the possibility that costly precaution may precipitate unwanted outcomes of equal magnitude. Indeed, some risk exposures may require substantial outlays to remedy them, or require significant curtailment of beneficial activities, thereby causing us to incur significant "opportunity costs". An example is global warming. There seems to be no easy way to curtail greenhouse gases. Nonetheless, some efforts may make sense in that they are easy and/or inexpensive to apply, yet still effective. This fact underlies many of our day-to-day, "common sense" applications of precaution. If I am crossing the street and see a tilted manhole cover in my path, I know that falling into the hole can cause grave physical harm, or even death. The probability of such an event is really of not much consequence (even if I could somehow "mentally" calculate it). Instead, I take a simple and effective precautionary measure: I step around the dubiously protected hole. Precautionists have incorporated this into a version of the precautionary principles cited above, by suggesting that costs be incurred to prevent loss as long as they are "reasonable".

Let's take an example of reasonable precaution from one of the more mundane, yet critically important, decisions in organizational and business risk management: Sprinkler installation. The functioning of sprinklers to protect a facility from fire, or to stop the spread of fire, is a very complex process. Often, sprinklers themselves cannot be depended on to "put out" a fire, but simply to forestall its progress until mobile firefighters and fire fighting apparatus can arrive. While insurance company statistics do show a difference in damage between fires involving sprinkler facilities vs. non-sprinklered facilities, many intervening factors complicate the analysis. Many buildings constructed before the modern era of fixed sprinkler protection are hard to retro fit. These buildings have other inherent hazards that increase the likelihood of fire. For example, they are usually built of frame timbers. They may also be "cheaper" properties, and therefore attract tenants who are less likely to afford other fire prevention improvements and processes. The processes themselves may be less than fire safe. All these factors make it difficult to assess the effectiveness of sprinkler systems from a probabilistic cost/benefit standpoint. So what? Sprinkler installation on new construction can be fairly inexpensive. One sprinkler manufacturer claims the cost per square foot of sprinkler installation is less than carpeting. Compared to the value of modern buildings, and the productive capacity they represent, sprinklers are fairly "cheap". A precautionary decision to install sprinklers may be an easy one. No complex (and perhaps, quite dubious) probability calculations need enter the decision, and they often don't. Insurer "sprinkler credits" on premiums encourage sprinkler installation, but are usually merely estimates based more on the desire to insure newer, better equipped properties (i.e., sprinkler credits rarely justify a major retrofit of a previously unsprinklered building). Cost/benefit analysis of sprinkler protection is the stuff of introductory texts on risk management, where probabilities of loss are given, and money spent or saved imaginary. In applying this modified precautionary approach, we do not "weigh" costs against benefits. Rather, we identify high-risk potentials and avoidance opportunities, and then proceed to determine if avoidance is "affordable".

What exactly constitutes "reasonable" is hard to define. It is definitely NOT based on cost/benefit, otherwise we fall back into that trap. It may not be an economic justification at all, but rather one based on social goals and responsibilities. The "reasonableness" approach requires that we maintain some proportionality between the risk we face and the expenditures we make to avoid them, both in terms of direct cost and "opportunity" costs. Suffice it to say for now that reasonable precautions do exist, which makes at least some precautionary actions fairly easy to apply.

3.5 Protecting human life

Initiatives for protecting public health and welfare, be they on an individual, business or community level, often demonstrate "reasonable" actions taken on a precautionary basis, applied to our most precious commodity: Human life. Scientific developments in medicine and study of danger of health effects of our every-day environment, both man-made and natural, as well as actions based on good old "common sense", assure that there is some level of protection offered us by actions that are to a great extent obvious in their effectiveness.

The cost of public health and safety initiatives is often measured using so-called "cost per life saved" metrics. The measurement of human life and well being in economic terms is controversial. As used in terms of proportionality, we can use this "cost" as simply a measure of monies spent to prevent. This number will provide a rough metric against which the "reasonableness", or alternatively the "expensiveness" of such costs may be compared. We do spend money to finance the eradication of catastrophic risk. When catastrophe itself is gauged in terms of a "life lost", we can compute the ratio of cost of safety (i.e., danger, or catastrophe, elimination) to lives saved. For example a recent study by the National Bureau of Economic

Table 3.1 The Costs of Various Public Health and Safety Actions

Action	Cost Per Life Saved ($)
Breast cancer detection	15,000
Auto seat belts	30,000
Aircraft cabin fire protection	100,000
Highway guard rails	100,000
Hypertension control	150,000
Children's sleepwear flammability requirements	800,000
Auto air bags	1,600,000
Land waste disposal regulations	3,500,000
Airborne benzene reduction	5,000,000
Signal arms on school buses	8,200,000

Research estimates that automobile seatbelt costs are approximately $30,000 per life saved – a modest amount by any standards. We do not use this metric to value human life, as in expected value cost/benefit. Rather, we use it as a cost indicator.

On the basis of this simple metric, it would seem that a variety of health and safety initiatives fit a rough classification of "inexpensive" efforts when compared to the magnitude of results (i.e., catastrophic). Table 3.1 shows some cost per life saved data associated with various health and safety actions, collected from a variety of official and unofficial sources. We see that there quite clearly exist reasonable actions given the exposure. (i.e., loss of human life).

The question becomes, were do we place the threshold of reasonableness? At $500,000 per life, $1,000,000, $5,000,000? By answering this question, we implicitly put a value on life. While such valuations seem distasteful, there must certainly be some threshold upon which paying is simply "too much". Eight million dollars seems like an expensive price tag for signal arms on school buses, but what if it is your child it saves? While we can make considerable progress in precaution without ever having to deal with the threshold head-on, simply that we are somehow below it, eventually expensive decisions will have to be made. We face, once again, the dilemma of precaution.

3.6 The importance of proper metrics

This discussion of precaution as it applies to the preservation of human life illustrates the importance of adequate metrics to decisions about catastrophic risk based on *any* sort of consideration of cost. Challenges to the effective application of practical precautionary actions can result from the difficulty of establishing costs with any degree of precision.

As we have shown above, we can select practical health and safety options based on cost per life saved. To do so, we establish a rough threshold of reasonableness. Some precautionary measures are easier, and more effective, than others. This makes them cheaper than others to apply. Underlying decision methodologies aside, costs can be one of the greatest differentiators between decisions. In expected value decision, how we value costs represents one side of the cost/benefit equation. How we value these costs clearly impacts the decision. Over-valuing costs has the effect of making expected value decisions *less* conservative. Preventive action or avoidance becomes more costly relative to the expected value of loss, and hence more options to eliminate risk are rejected in favor of risk acceptance. In terms of the proportionality of precautionary costs, exaggerated costs can make precaution seem more "expensive" than it really is, and *vice versa*. For example, asbestos is a material that has proven carcinogenic characteristics. How we protect ourselves from these characteristics is a matter of debate. Establishing limits for occupational exposures can be done relatively cheaply, on a cost per life saved basis. Cost of removing asbestos from existing installations however carries a cost estimated by some to be as high as \$100,000,000 per life saved. In this way, the costs of various preventive actions can themselves become controversial.

Obviously, assessing costs adequately is an essential precondition to identifying those that are "reasonable". So while modifying the application of precaution from its pure basis, "avoid (possible) catastrophe", to that based on reasonableness, "avoid (possible) catastrophe *at reasonable cost*", we have reintroduced issues of cost measurement and balance that often plague cost/benefit analysis.

3.7 Reasonable precautions and human evolution

Identifying a potential risk as something requiring precaution is therefore similar to any other process where the human being recognizes a compound, or complex concept. It's a form of *pattern recognition*: We observe certain patterns in data, and act accordingly. The process is different than action based on precise measurement of properties, and subsequent calculation. We don't so much measure risk as recognize it. The application of precaution can be thought of as representing a very basic form of human reasoning. We recognize something as food, and eat it. We recognize something else as shelter, and take cover within it. We recognize something as dangerous, and take precautionary action.

Precautionary principles are evident in human "rules of thumb" for dealing with danger. "Safety first", "better safe than sorry", "an ounce of prevention is worth a pound of cure".... These colloquialisms embody the simple, practical rules of precaution by which we lead our lives. These all suggest that as individuals we are fully aware of the importance of precaution, have the ability to at least roughly discern dangerous situations from those that are not, and are able to take actions that are an effective counterbalance to these dangers.

It is arguable that such risk recognition is a basic "instinct" in human beings, much like the ability to optimize adaptively when conditions permit. That some element of precaution seems logically necessary to human survival, combined with a rather remarkable streak of existence on this planet, suggests that practical elements of precaution have existed throughout the history of the human being. The ability to act with precaution would seem like a precondition of effective evolution. Interestingly, many plants and animals share the same history of existence on this earth, some actually existing far before humankind. Undoubtedly, these species share many adaptive features of humans when it comes to optimization in their respective environments. We might expect as well that simple pattern recognition of dangerous situations might guide these species as well. A fascinating and potentially very instructive line of investigation opens up: How do other species on this earth "manage" risk?

While evolutionary history suggests that practical application of precaution may have gotten us to where we are today, can it assure our future survival? The question becomes, are precautionary dilemmas becoming more common as the world progresses?

3.8 Facing the limits of practicality

There are reasonable precautionary approaches that are safe, simple and relatively "cheap". We looked at many of them here, and if we put our minds to it, we could probably come up with a host of common risk management strategies that rely on such forms of prudent avoidance at low cost.

As the world becomes more complex, however, the challenges mount. It becomes difficult to tell if some expensive precaution might be the one to prevent the next catastrophe. In hindsight, as long as the preventive measure cost less than the catastrophe, it can't be unreasonable. Similar problems occur when we attempt to "prioritize" our approach to catastrophe. That is, treat some, perhaps the "larger" ones, first. Of course the whole idea of "relative catastrophes" doesn't seem to make all that much sense: When you are ruined, you're ruined. Irreversible effects are just that, there is no "going back". Catastrophe is more a threshold than a continuous measurement. It is while we are working on preventing catastrophes #1 and #2, that #3 will get us (Murphy's Law, liberally applied). It would seem that, logically, catastrophe avoidance is an all or nothing thing. And as exposures multiply throughout our evolutionary history, the problem grows greater. This all just compounds the dilemma of cost. There is the potential that small incremental costs can "creep up" to large total costs as we apply precaution consistently over a widening field of risks.

Clearly we must view the effects of precaution globally, not locally. That is, consistent application of the principle requires that we consider its application to the variety of risks we face, not one by one, individually. The problem is that while each decision may look fine when assessed by precautionary criteria, in the application to multiple risks, the expense will overwhelm us. The problem of local versus global application is not unique to precautionary criteria, however. How we frame the decision can affect the results of other decision criteria as well. In applying expected value decision-making we must also be aware of the wider landscape of risk, lest "wrong" decision result. For example, a firm may assess the expected cost potential of windstorm, fire and product failure independently, settling on an optimal mix of loss prevention for each. However, the analysis may not include the potential of various events happening during the course of the year. For example, a flood and a catastrophic product failure may occur in the same year (as independent events) thereby pushing the total cost of loss up far beyond what may be considered on an individual basis.

On a societal level, risks can mount in a similar fashion. Technological innovations may proceed on various fronts, each with their own set of special risks. Population growth itself may engender further catastrophic potentials that require precautionary action. For example, land usage may expand into areas that are prone to natural disasters such as flood or earthquake, challenging our ability to protect the people who live in these areas. This all suggests that precaution must be examined within the wider scope of our policies toward human evolution and progress.

4

Precaution and Progress: Identifying Alternatives

The consistent application of precaution may lead to the dilemma of unacceptable expense, and eventual impasse (the "dilemma of precaution"). Reasonable precautions make sense, but limiting our precautionary approach in this fashion does not solve the catastrophe problem. A more far-reaching approach to precaution is needed. This approach requires a broader view of progress that depends on analyzing the potential paths of progress (be it economic, scientific or social) with an eye toward possible risks. It makes sense that genuinely effective precautions must occur *before* the potential for loss is realized. Applied in this manner, precaution may eliminate the activity before its creation leads to a paradoxical position in which we are essentially "stuck" with the activity, for better or for worse. Precaution in this sense is "anti-progress" only to the extent we have already made irreversible progress toward possible disaster.

4.1 A wider view of planning

Consider an example that involves business and community risk management in the face of natural perils such as windstorm and flood. While statistics on floods and windstorm exist, the time span through natural history that we can observe such events is fairly narrow. Maybe 100 or 200 years or so for reliable data. The probability information so obtained is very uncertain. In addition, the potential damage to persons and property can be immense. Nonetheless, reasonable precautionary approaches can be taken with respect to building facilities and homes in such areas. Where the potential exists for truly damaging windstorms and floods, these areas should be avoided. Note that precautionary actions appear unreasonable or unworkable here only after the decision has been made to construct in flood or wind prone areas. Then what do we do? No amount of probability data can help us come up with the solution. We either stay put and hope for the best (the fatalistic approach), or move out and incur the considerable costs of abandoning already constructed properties (expensive, post-fact precaution). Pooling, *via* insurance, can provide some relief from economic losses, but only within the realm where these losses occur with sufficient regularity that they can be treated statistically. The insurance mechanism does not properly respond to non-economic losses, such as loss of life. Even from the standpoint of pure economic losses, the insurance mechanism can be easily overwhelmed by losses of sufficient magnitude.

It is clear that such wider views of planning, or should we say pre-planning, are rarely taken. More usually risk managers are tasked with finding solutions once the technical parameters of the risk have already been established. It is in these cases that many find the precautionary approach unworkable, and hence tend to generalize the idea as useless. We set up the preconditions for this failure by not employing precaution earlier on in the overall planning process.

Applying precaution earlier on in the process is a solution that requires a closer look, and possible redefinition, of what we consider "progress". That in turn requires a more careful appraisal of how we fit into our environment, our society, and indeed the world itself. The challenges here could themselves be enormous. However, the promise of such re-evaluations is huge as well.

Risk follows progress. This means risk is a temporal phenomenon. This dynamic aspect of risk is rarely recognized. We proceed on a path of technological progress, and stop periodically to assess the risk associated with this progress, at a point in time. By assessing risk at these various points, we are unable to address the dynamic processes in between, those that actually lead to increased risk levels. Periodic evaluation suggests an approach that whatever is done is done: Now what do we do about the risk?

Figure 4.1 is a dynamic version of the simple risk management decision matrix shown in Figure 2.2. The cost of preventive action is net of opportunity cost, or foregone benefits (as shown in the extended decision matrix in Figure 2.5), as well as "sunk cost". Sunk costs relate to various infrastructures that support the beneficial activity. These could be a petroleum industry that supports hydrocarbon-fueled automobile, for example. A large part of the investment in this infrastructure becomes obsolete if the fuel source for autos is replaced for precaution-

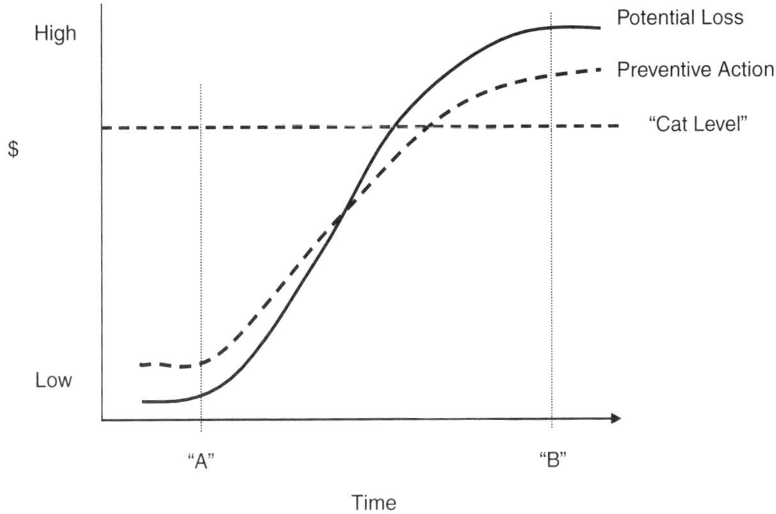

Figure 4.1 A Dynamic View of Risk

ary reasons. Likewise, New York City, one of the world's largest population centers, is vulnerable to catastrophe due to its coastal location. The cost of "moving" the city inland would be absolutely enormous (so, of course, would any potential catastrophic event).

This preventive cost is the number that we put in the "Take Preventive Action" cells of the risk decision matrix. The potential loss is the number we put in the "Do nothing/Loss Event Occurs". Also shown is the level that entails catastrophe ("cat level"). The chart therefore shows how these cells, and hence the decision problem, changes over time. While we show the potential loss and benefits crossing at some level, the fact they do has no impact on our dynamic analysis. Our concern shall be only that they eventually both cross the catastrophe level, and hence the dilemma of precaution ensues.

The presumption is that we can suitably adjust the risk to some acceptable level as time goes on. That is, we progress from point A to point B, stopping at B to assess risk. Simplistically, we prepare a risk matrix, and apply the minimax rule. Where the cost of mitigation and any possible alteration or curtailment of progress remain "reasonable", we can exercise these simple low cost actions to mitigate catastrophe risk. As long as risk remains manageable along this path, i.e., it doesn't get "out of control", periodic assessment is fine.

The problem comes in when the periodic assessment all of a sudden presents us with a cost of risk avoidance that is high compared to the gains progress has achieved for us. We prepare the risk matrix at point "B", and find now that we face the dilemma of precaution. If we curtail the activity, we lose the benefits, and quite possibly have to incur added cost, due to the development of an "infrastructure" around this progress which now becomes unusable, or which might even have to be dismantled. While there may be an infinite number of pathways between point A and B in a scenario such as that depicted in Figure 4.1, what we end up with is a dilemma that is only realized after some significant period of time passes. Within that time, the risk-causing activity entrenches itself. At point B, precaution becomes difficult, and perhaps even unworkable (in terms of the dilemma). If we had considered risk as part of the planning process, from the beginning, the dilemma may have been avoided.

It is logical that the faster technological development proceeds, the greater the chance that risk goes out of control between monitor points A and B. As it stands to reason that technological progress is proceeding today at an all time rapid pace, we should become more concerned with the prospects for risk dilemmas arising.

The rise of risk potential between points A and B can also be subtle. One of the most insidious varieties of risk is "creeping risk", or risk that slowly accumulates over time. Here the danger is that it may not be noticed until it is too late. By "too late" we mean that point at which both the catastrophic potential and the costs to eliminate both become high (the "dilemma"). Both heart disease and cancer have been suggested as being creeping risks that have achieved potentially (if not actually) catastrophic levels. Both have been associated with an increasing level of industrialization that expose us to unprecedented levels of complex chemicals with unknown or little known effects on the human body, as well as a host of environmental stresses. The elimination of the catastrophic potential would require a substantial "restructure" of the way we live.

This all does not mean that Figure 4.1 represents the typical path of risk. Again, an infinite number of paths can occur between point A and point B. What we are trying to point out is that a path such as that shown in the figure is how risk dilemmas can, and probably do, develop over time. The risk decision matrix is dynamic. We can start with one that presents a fairly "easy" decision with respect to the application of precaution. We can end up, over time, with one that represents dilemma. Nor does this discussion suggest that proactive risk management is simply a matter of monitoring risk from point A to point B. Proactive risk management starts before we ever undertake the activity in question. What we try to do is anticipate the path of risk between point A, point B and beyond.

4.2 Assessing alternatives

Avoiding precautionary dilemmas means getting "ahead of the curve" that leads to catastrophe and related dilemmas of prevention. We need to be able to identify future scenarios for progress, and their associated failure points. Particular attention must be paid to the "worst case" scenario(s). Unlike our approach to statistical risk, we need to make decisions related to high-stakes risk as if we won't get a second chance. This means that any iterative approach that embodies a wait-and-see attitude is unacceptable in the high-stakes domain.

Alternatives assessment is a formal approach to the pre-assessment of risk that requires that all actions be evaluated with an eye towards future risk scenarios, or paths, through time. This approach requires as much or more scientific thinking than "periodic" assessment and the attempt to solve risk dilemmas once they become entrenched. If the assessment of alternatives is made an integral part of planning progress, the potential for dilemmas is greatly reduced, or even eliminated.

We could believe that the path from point A to point B has tremendous positive potential associated with it. Simultaneous assessment of the situation could indicate that at point B, we can expect significant, possibly catastrophic, risk. Can we get from point A to more advantageous point B using some reasonable alternatives? *Dynamic risk assessment triggers the search for alternatives.* Therein lies its value.

Under alternatives-based precaution, high expenses and the dilemma of precaution become a call to action, not an excuse for inaction. It is these risks that demand our greatest degree of effort and scientific creativity. If the cost per life saved becomes "unreasonably" high for an activity, we seek to reduce that cost. If, based on some expected value calculation of costs versus benefits, we reject precaution, we have forever forestalled further investigation into mitigation. While some argue that precaution forestalls scientific progress, it is expected value decisions that can have the greatest negative effect on scientific progress, if the progress we value most depends on a safe and secure future. Science must not proceed to expected value standards, it must proceed to precautionary standards. The perceived strictness of precautionary standards is only when compared to the lax standards of expected value. In practicing precaution we set the ultimate standard: A safe world. The truly *un*scientific approach is to say that this world is not achievable, and we must simply live with that fact.

Nothing here suggests that alternatives assessment is easy, or that the "right" path around precautionary dilemmas will some how magically make itself known to us. We would offer at least the suggestion, however, that assessing the potential for such dilemmas before they occur may (1) present us with a wider array of viable (workable) solutions and (2) permit us to make some tough decisions, early on, about whether such risk potentials be allowed to manifest themselves. That is, if we are to become fatalists, we should at least be allowed to do so voluntarily, and not be forced into the position.

The most egregious form of precautionary dilemma is that which could have been prevented by utilizing an available alternative. We don't mean from the standpoint of hindsight, but rather by prudent application of alternatives assessment early on in the process of planning for progress. Mistakes, errors, bad information, poor execution, these can all derail the most well-intended assessment of alternatives. The point is, at least we tried. The measure of success in such attempts will likewise be the measure of the effectiveness of our science. Again, *it does not make sense to put faith in a science that promises progress, only to fail at providing us a degree of security in that progress.*

4.3 An illustrative example

As an example of proactive planning within a manufacturing organization, let's examine the concept of *clean production*. Critical decisions on the progress of the firms, along some potential "growth path", are often dependent on choosing appropriate production technologies. The wastes produced by these technologies create a financial burden on the company, as well as possible ecological burdens on the community. Clean production suggests that development of appropriate production technologies must consider the potentials for serious risk as part of the planning process. For example, the machining of metal parts requires the use of various cutting oils for lubrication. Once the cutting and shaping operations are performed, the parts must be cleaned, using a variety of solvents. Considering the entire production process the proactive, precautionary production manager may consider the use of biologically based cleaning agents that have a relatively short "after life" in the general environment. That is, they breakdown, or *degrade*, with little or no adverse effect on the general environment. These products represent viable alternatives to petrochemical-based solvents that can have serious adverse effects on both the environment and people. And once a production technology is built around a particular cleaning process, the switch can be costly. Avoiding possible catastrophic environmental and human health effects then becomes subject to the dilemma of precaution, in terms of the high costs to remedy the problem.

Modern pollution control efforts that rely on "end of pipe" treatments are an example of attempts to manage the risk of an exposure, once the infrastructure surrounding the exposure has already been entrenched. At some "point B" in the history of industrial production, it was realized that the waste byproducts of these production methods resulted in a substantial amount of environmental pollution. Such pollution has the potential for catastrophic effects on the natural environment, human beings included. The risk management "fix" was the installation of pollution control devices to clean up resulting wastes. While not to disparage the tremendous progress that has been made over the years in the reduction of pollutants that enter the environment, via the utilization of pollution control technologies, such control has been expensive, and often less than completely effective. Pollution control in this "end of pipe" fashion is often pointed to as a failure of precaution. Yes pollution has been reduced, but at a high cost. The potential in some industries is that they may be driven out of business by the cost of pollution control. The answer is clearly not to relax restrictions on how much pollution eventually enters the environment. Resorting to cost/benefit in terms of expected value calculations does not provide credible guidance, and, due to the uncertainties involved, is subject to considerable manipulation by self-interested parties. On the other hand, sensible, affordable precaution comes in the form of assessing production alternatives that reduce waste within the production, use and disposal process. Control of the disposal of non-degradable plastic packaging helps control environmental pollution, but it is expensive. Clean production suggests the development of biologically neutral packaging is more effective at pollution control (and that means, catastrophe protection), and ultimately be cheaper.

Clean production initiatives show that a wider view of precautionary risk management, centered within the overall planning process can save money as well as avoid the dilemmas associated with the inevitability of possible harm. This places alternatives assessment at the center of a proactive risk management approach.

4.4 Natural vs. man-made risk

Planning for progress requires some sort of baseline by which this progress can be measured, as well as a goal for which we can strive. A beginning and end, if you will, which circumscribe our journey. In planning for progress, we need to identify our overall risk goals as well. In the dynamic view of risk, as an integral part of planning for progress, we need to assess what ideals exist.

It is often noted that those who apply precautionary risk management approaches assume a benevolence of nature with respect to risk and harm. This is not a surprising feature of precaution. It does not, as some critics believe, show any sort of bias on the part those taking a precautionary stance. Rather, an unfettered natural environment provides an ideal "background" level for risk. In terms of progress, our goal may be to maintain at least this natural level and improve on it as we may. For example, in a natural setting, humankind may face the perils of exposure to the natural elements – sun, wind and rain. An improvement in our condition, including a reduction of risk, would inure from us devising some sort of shelter, or protected living areas. Housing, buildings and other dwellings provide this protection for us. By clearing land and building housing, we have in the process altered our natural relationship to the outside environment. The risk associated with more routine interactions, can be assessed statistically. Do these actions result in any catastrophic perils, the existence of which would need to be treated with precaution. If so, how would these precautionary treatments be reconciled with growth? In this assessment process, we build a dynamism that is designed to maintain a balance with the natural environment.

The precautionist need not argue here that nature is completely benign. He or she need only point to the fact that the balance must be maintained. That progress is needed in order to reduce or eliminate natural perils is obvious. That progress would proceed by ignoring these perils is ridiculous. That progress must be obtained at the cost of constant imbalance is also an untenable position that does not jibe with reality. The presumption is the natural environment, with minimal intervention by humans, has sustained human life for millions of years.

When we speak of nature, thoughts immediately turn to our physical environment. The idea of "naturalness" however must go beyond the merely physical. There is a natural view of economic interactions, social ones, political and even psychological (i.e., "human nature").

These may be more difficult to define, and even controversial at times, but it is reasonable that natural guidelines exist in these areas of human existence as well. Our actions in this area have a similar motivation than in the physical arena. While our focus is hazard risk, we mention these here because they can have some interface with how we treat the more obvious physical risks.

Critics of precaution sometimes argue that the naturalistic approach is anti-progress or anti-technology. This of course implies that technology and progress are incompatible with natural harmony. To this extent, the critics have tipped their hand. Nothing implies that there is an antithesis. What criticism against a naturally precautionary stance implies is that there is an agenda to protect *special interests*. And that agenda involves a decidedly anti-nature stance, prompted by the fact that the destruction of nature can lead to short-term profits by those that engage in it. Criticisms of self-interest or bias on the part of those that suggest a natural precaution often reflect the opposite bias.

Assuming that nature is malevolent and needs to be tamed *via* technological development, is an extreme and ultimately untenable position. There is no evidence that natural forces are afoot that aim toward the natural extinction of humankind. In fact, almost all the risks that can be objectively cataloged today are of the "man made" variety. Quite a few of those revolve around the creature comforts, rather than genuine survival. The effects of many of the natural perils that exist, such as earthquakes, windstorms and floods, are exacerbated by the conscious activities of man (such as land use policies).

This suggests that the notion of man-made exposures extends beyond the creation of technological exposures. Decisions on how we behave as an individual, business entity or society toward natural perils is a crucial part of the risk management process. It is human kinds interaction with these natural perils to which precaution applies. In some cases it is hard to see how changes in the way we live our lives would affect something like the possibility of a large asteroid hitting the earth. There the exposure is *part* of the natural environment. Precautionary decision-making in this case is based solely on the potential for such catastrophes. Certain natural perils have existed since the beginning of time, and are part of the natural background risk. The existence of some level of background risk is all the more reason we should not increase this already existing level via man-made exposures. By building in flood zones we are taking human-made actions that increase the potential for disaster from natural perils. The extent that such activity is part of progress needs to be carefully assessed, as do the alternatives.

The idea of an idyllic natural environment, a Garden of Eden if you will – everything in physical, social and economic harmony – has certainly been romanticized to some extent. Yet, when we need a vacation from the hustle and bustle of the day-to-day world, we don't visit a robotic manufacturing plant. We head for the mountains, or the ocean, or the lake. The natural environment, idealized or not, still represents a certain harmony in the life of humankind which serves as a reasonable benchmark for the "good life", and should hence present a natural model for what precaution is meant to preserve.

Last but not least, natural risk represents a very honest approach to risk. There are no issues of trust. Natural risk is equitable – fair. No one is "taken advantage of" under natural risk. No special interest to influence decisions, either consciously or unconsciously.

While we can't say for sure that a "natural" benchmark is the most appropriate, a lot of things suggest that it is. It makes sense to use it until, or if, something more appropriate comes along. At the very least, the natural approach deserves its own promoters. Those that would promote a simplistic regime based on unfettered technological "progress", as measured by say, "output/income growth" (personal wealth, net profit, GDP), are well-funded and equipped by their benefactors: The interests that promote such growth as part of their own plan of progress. Who fights for the "natural" position? Clearly nature itself is in no position to do so.

One need not have a bias toward natural living, or an overtly "environmentalist" stance to want to know the answer in order to make sure that natural position is well represented. Otherwise, we could find ourselves being overwhelmed by the opposite bias. From the standpoint of our physical environment: If a logging company seeks to farm public land we need to know the impacts. It is not unduly suspicious position to think that the logging company may not be able to offer a perfectly unbiased assessment of possible adverse consequences. So who will? Who will assess the direct impact on me, my children and my children's children? It is in fact a demonstration of a lack of bias toward any position to actively seek equal advocacy for all of them. Making sure the natural position is well represented should be a goal of anyone who dislikes the idea of being taken advantage of.

In the United States justice system, when an accused cannot afford counsel, one is appointed. These "public defenders" assure that everyone gets a fair trial, regardless of their ability to defend themselves, or pay for such a defense. To a great extent, the regulatory structures of many governments around the world provide a defense of the natural position of risk with respect to progress. As we will see in the next chapter, more and more regulation of catastrophic risk potentials around the world is being based on precautionary standards.

4.5 Shifting the burden of proof

In the practical application of precaution, the primacy of natural/humanistic position with respect to risk manifests itself in a shift in the burden of proof that a new or existing exposure is indeed "safe" to those that would propose the activity. In terms of our assessment of the possibility of catastrophe, we assume the widest interval of uncertainty consistent with our knowledge level when assessing a new exposure. If that interval extends into the precautionary region, so be it. The burden of proving that the exposure is sufficiently safe (i.e., sufficiently low, or not, overlap with the precautionary region) is left to those that endorse the exposure.

For example, there are those that while respecting the primacy of the natural stance to the physical environment, suggest that this environment exhibits a certain "assimilative capacity" towards some man made elements that might otherwise considered detrimental, and possibly catastrophic, pollution. To some extent, a product or activity may be considered justified on the basis of this assimilative capacity to reduce the likelihood of serious damage to the environment below some de minimis probability level. This fact must be demonstrated, however, and not taken for granted. We know, for example, that our early assumptions about the physical environment's capacity to assimilate the effects of chlorofluorocarbons (CFCs) were incorrect. There is significant evidence that CFCs can damage our environment by depleting the earth's protective ozone layer of the atmosphere. CFCs have since been the subject of regulation on a precautionary basis.

Among the further criteria for burden of proof: Have the alternatives been evaluated? The question is not "do you have (or will you have) the technology to make the exposures safe into the future (say, at "point B")? The question is rather, "can the exposure prove itself the safest of all the alternatives available". Burden of proof requirements are therefore a further impetus to the careful assessment of alternatives. We do not select, say, the "least costly" alternative and then assume it is safe until proven otherwise. If it is the least costly alternative we propose, we need to prove it is a safe alternative as well.

Precautionary alternatives assessment requires that we recognize the potential dilemma of "point B" in Figure 4.1 *before* we engage in the activity. This can introduce a sort of precognitive version the precautionary dilemma. On the one hand, this anticipatory dilemma is less onerous from the standpoint of catastrophe as we have not yet induced the exposure to loss. The situation is utterly safe, as we have not engaged in the action. However, by having to "prove" the activity is safe (using an accepted "possibility of harm" threshold), we forestall *future* benefits.

To support the claim that precautionary burden of proof can forestall future benefits, critics often take a backward look in time, applying precaution to activities that some would consider "models of progress". They make the assumption that such progress might not have been made had precaution been applied. Most of these examples reveal failures of the imagination to perceive effective alternatives, rather than failures of the principle itself. Where would we be today without the automobile? This pronouncement assumes we mean the ubiquitous auto of today, powered by the internal combustion engine. If we narrow the question to mean where would we be without internal combustion engines that rely on hydrocarbon fuels, the answer might very well be a whole lot healthier and safer. We should not confuse ease of transportation – a critical and beneficial level of progress above our natural state (transportation by foot) – with the specific methods of transportation at hand. We do so only because we believe, and indeed, led to believe that what is at hand is the "only way", or the "best way". This suggestion has come to us via a process of evolution. However, it is a process of evolution that has, for the most part, ignored dynamic risk assessment.

What if precautionary evolution of personal transportation had proceeded on the basis of "clean" energy sources, such as electric motors, possibly solar powered? Much of the history of internal combustion engine geared towards hydrocarbon fuels is based on fortuitous events, such as the availability of relatively cheap sources of oil. Innovation is also stunted by the development of powerful commercial interests, i.e., those that develop a "stake" in a particular technology *via* large, immobile investments.

As we have suggested above, everything is not risky. There are benefits to many activities that can be realized safely. There are other proposed activities that can be quite readily assessed as creating unacceptable risk, and hence dismissed. It is those "in between", the ones that entail a level of uncertainty as to the possibility of catastrophic potential, that create the great difficulties for decision. The fact is that it is more likely that "new" technologies or endeavors will entail the greatest uncertainty, and hence face the highest hurdles in term of proof of safety. But isn't this the way it should be? The uncritical acceptance of a new technology, might not only create the possibility of disaster further on in its history, but the costs of later abandoning this technology may themselves grow to tremendous proportions. At that point in time, we face the dilemma of high-cost precaution.

Once again, there is nothing inherent in the concept of precaution that prohibits the assessment of the risk of precaution itself. To the extent that credible benefits from a new technology can be identified, the opportunity cost of precautionary avoidance should be considered. All costs considered, we might face the precautionary cost dilemma. Those that would suggest that we go forward with the new technology at this point are clearly what we would call fatalists. The identified benefits of the proposed technology or activity make going forward with the process (i.e., "do not avoid") the least cost alternative (in terms of preventive outlays). It is not the case that "benefits" have exceeded "costs" – they haven't. That balance is irrelevant. The operative decision philosophy is based on the fact that a dilemma exists, and we can't do anything about it, so why worry? Very simply: We again find ourselves in the position of being doomed if we do, doomed if we don't. If we "do", we're fatalists. If we "don't", we're precautionists. Whatever the choice, be prepared to live with that philosophy.

4.6 Alternatives assessment vs. post-fact risk management

Most business corporations have groups within the organization that are responsible, ostensibly, for "risk management". It is not uncommon in larger organizations to find someone with the title Risk Manager. The endeavors of risk management departments of large corporations reinforce a static approach to risk. Exposures are assessed periodically using "surveys" of potential risks. These surveys serve as nothing more than a catalog of outcomes that may result from an organization's technical structure at any point in time. How many buildings does the business own? Is the business near a river? Is there a potential for product recall? This so-called exposure analysis, or assessment, remains completely static. It is applied periodically, sometimes yearly, most often after several years, or more. The results are often used to adjust insurance programs. For example, if we have just opened up a plant near a river with a flooding potential, don't forget to add flood insurance to our property insurance regime.

Organizational risk management often proceeds on the basis of a sequential model. The model can be summarized as: identify exposures to risk (exposure analysis), assess (i.e., determine) their probability/loss characteristics, and take appropriate action *via* matching of risk management treatment to the risk characteristics (Figure 4.2). This last step involves selecting the most cost-effective (i.e., cheapest) protection option (often, some form of commercial insurance). The results of the exercise are then periodically monitored for effectiveness. This "I-A-T model" (identify-assess-treat) is ideally suited to the management of statistical risk, as any deficiencies in the approach can be corrected iteratively, through time, based on observed statistical performance of the plan of action. The only consideration of "alternatives" in the I-A-T model, however, occurs in the "treat risk" stage, were the risk manager selects from alternative statistical risk treatments (including insurance), based on their "cost effectiveness".

Usually, there is little specific guidance on the I-A-T process given by senior management. Rather, guidance is sought from management in the form of company objectives with regard to physical and financial operations: "What is our corporate plan over the next year?". The risk

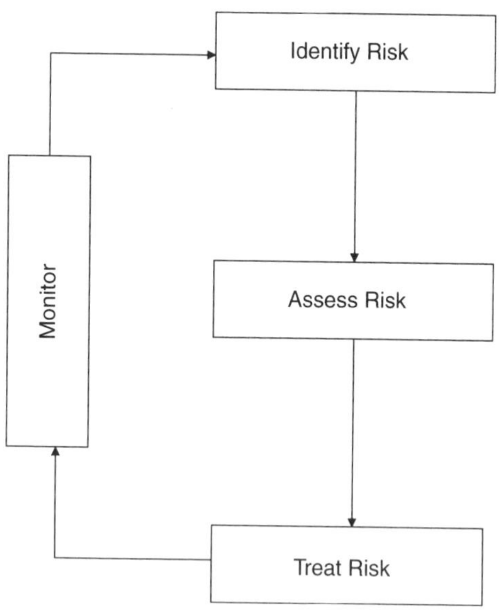

Figure 4.2 The "I-A-T Model" of Risk Management

manager identifies any new exposures which may come into existence over the next planning period, as well as potential changes to existing exposures, assesses their effect on the firm's statistical risk profile, and plans for appropriate action. This is essentially the I-A-T model at work. Insurance programs and loss prevention programs are adjusted accordingly. The process is completely reactive. We might call this application of I-A-T the *post-fact approach* to risk management. Risk management plans are established *after* exposures are set.

Performance monitoring of the I-A-T is accomplished *via* simple "cost of risk" metrics for assessing risk management performance. Cost of risk, as most commonly defined, is nothing more than the sum of statistical losses (most, "self-insured" *via* some sort of deductible to an otherwise fully insured policy) and insurance premiums. The basis for the assessment of risk management performance under this regime is, therefore, purely statistical. Considerations of catastrophic potentials only enter the process tangentially, usually in the setting of insurance policy limits. There as well, the analysis proceeds on the basis of given exposures, and is often assessed based on the limits of companies in similar businesses ("benchmarks"). Proactive assessment of loss potentials is rarely considered, and almost never for "uninsurable" perils, such as failure of a product to perform as expected ("warranty losses") or market failures. The idea that insurance premiums fully internalize the costs of accidental losses is seriously inadequate as a basis for risk policy.

As a result of this reactive, post-fact approach to risk, alternatives assessment in the wider sense is ignored. Alternate exposures are not considered, only alternate treatments to existing exposures. The problems of expensive post-fact treatment of risk via precautionary approaches surface, including the prospect of expensive treatments for entrenched risks (i.e., precautionary dilemmas).

Some organizational risk managers have espoused a wider approach to risk that goes beyond the risk of physical hazards and their effects. This approach forms the basis of so-called *enterprise risk management*, or ERM. It extends the explicit management of risk to economic risks, including those arising from financial operations, financing and investments (including credit risk), and strategic risk with respect to the market, as well as the economy in general. While touted as a "holistic" approach to, the way ERM is applied today often falls short of being effective in the management of truly significant risk. For one thing, ERM has a distinctly internal focus. That is, its concentration is on economic risks that impact the firm's internal functioning, without explicit regard to how this functioning affects the outside world. To some extent this is understandable, as in most cases, the individual firm may have a relatively small impact on the wider economic environment in which it operates. ERM is also more of a technical approach to risk, rather than a genuine philosophy of risk. Most of its benefit is based on the consolidation of like statistical risks, and the recognition of potential correlations among statistical risk that could make independent management inefficient. Even on this strictly technical basis, the value of ERM (compared to the independent treatment of hazard and economic risk) is often questioned.

To be truly holistic, "risk policy" must consider the effects of risk on all members of the society, as well as the global society we are all effectively a part of. This outward looking view of risk is very different than the one we take today. Most formal aspects of decision with respect to significant risk still rely on scientific looking estimates based on expected value criteria. They often fail to include the wider picture in their "cost" calculations.

For most efforts within the modern business entity, "doing the right thing" amounts to maximizing profit for investors. Profit maximization is a proper goal for insurance management (with respect to hazard risk) and statistical ERM (encompassing economic risk). The contribution of risk management to the effort is measured by the simple "cost of risk" measure – insurance premiums, plus statistical losses. Obviously, the lower this cost, the better. Profit maximization is *not* an appropriate goal, or not at least the sole goal, for a truly global view of risk management. The risks generated by the average business entity entail too many externalities. These external costs are seldom internalized as part of the standard accounting process, and hence may undervalue the efforts of a global view of risk management. This failure to account for externalities is what often relegates the role of risk management of the business enterprise to the management of statistical risks engendered by traditional hazards, such as fires, windstorms and floods. It is this failure to recognize the global nature of enterprise risk, along with the inapplicability of many of the statistical methods for dealing with risk, which will ultimately stall the progress of any enterprise-wide view of risk that relies only on post-fact risk management.

Will senior management invite into the setting of corporate risk policy the same individual whose primary concern, up to now, has been managing slips and falls in the company parking lot, and getting the best "deal" on insurance premiums? We may apply a sophisticated title to those individuals (Chief Risk Officer, or CRO, if you will), but the fact remains that if their skills remain bound to the statistical end of loss (the "tip of the iceberg"), they have little to add to the wider process of setting risk policy. The comfort the community, or the enterprise itself, gains from such efforts is small.

Closely allied to the financial management of risk in the modern organization, are the efforts of safety experts who deal with the "engineered" aspects of risk, as part of their operational (production) functions. These experts may specialize in the integrity of products and services offered to the general public, the safety of operations with respect to the general public, and safety of those that are employed by the organization. There are also those with the responsibility for the integrity of facilities and property, in the face of a variety of natural hazards, such as fire, wind and flood, as well as man-made threats to property, such as theft, vandalism or terrorism. Again, much of this endeavor proceeds on the basis of statistical analysis, following the I-A-T model of risk management. The functionality and safety of products is often assessed in this fashion, with the safety expert collecting data on mishaps, analyzing it, and providing cost-effective solutions. Likewise, work place safety experts gather information on statistical losses in the workplace and suggest safety improvements. If during the course of a month, safety records indicate a significant frequency of cuts to the hand, a cost effective safety improvement that includes cut-proof gloves and additional training may be in order. These analyses proceed on a statistical basis, and usually involve relatively "low" risks. High-stakes issues, once again, fall outside the realm of statistical analysis. Some cues are taken from the assessment of statistics on a wider basis, thorough the analysis of multi-organization data, collected and provided by a trade association or public entity.

Safety requirements based on a wider statistical analysis may also form the basis of regulation of safety matters by governmental entities. Safety regulations at the public level are often based on statistical cost/benefit assessments. The firm's sole component of risk assessment with regard to important public safety issues may be simply a matter of regulatory compliance. The "risk" is that they may face sanctions for violating these regulations, or that new regulations may find them unprepared to implement them. A "risk policy" based on statistical analysis and compliance with government safety regulations is valuable to the business and the community. It does not constitute a risk philosophy to the extent it excludes consideration of high-stakes risk, beyond the realm of statistical analysis. Like insurance, compliance with safety regulations is primarily reactive. Alternatives assessment is only considered to the extent that the decision is driven by statistical cost/benefit, albeit on a scale beyond that of the individual firm.

The extent to which safety experts participate in the wider arena of company risk management is usually limited to their interface with organizational risk managers, whose focus, as we have already suggested is on the insurance mechanism. Safety experts help make the case, usually statistically, for insurability and perhaps some premium reductions, based on anticipated future statistical performance under the insurance policy. Insurance also provides a "safety-net" that while seldom explicitly recognized, influences organizational safety efforts: Let's do what we can for safety, but make sure an adequate insurance policy is in place as well. With regard to the introduction of new products and services, the risk management approach is distinctly post-market. Any adverse effects of these products are treated in a post-fact fashion, after they cause problems. With regard to those issues that manifest themselves statistically this approach, again, makes perfect sense. For catastrophic risks, it practically guarantees that precautionary dilemmas will result.

With regard to the safety or production and its byproducts, the risk management approach is also often post-fact. As we have noted above, the post-production approach to environmental pollution, *via* end-of-pipe treatments, is clearly reactive. While they may achieve results that forestall or even eliminate the serious adverse effects of environmental pollution, they do so only at great expense to all involved. A constant struggle between the cost effectiveness of such regulations and the need to protect the environment results in these essentially reactionary treatment policies being seen as an ineffective solution to the problem. The example of clean production, described in detail in Section 4.3 above, shows how precautionary alternatives assessment can proceed to set the firm, and its community, on a course that steers clear of catastrophic potentials. Such efforts are the result of an explicitly precautionary philosophy of risk that embraces alternatives assessment to achieve its goals cost effectively.

If the sum total of a businesses explicit approach to risk is contained in a risk policy designed to deal with statistical losses, combined with a plan of regulatory compliance policy and commercial insurance for the "unforeseen", then that entities risk philosophy is incomplete. It may in fact be the case that philosophy with respect to high-stakes risk is implicit, or non-formal. In most such cases, an informal philosophy of high-stakes risk tends to be fatalistic. "Take charge" actions are formalized, and indeed glorified, in the form of explicit policies and procedures. Fatalism, on the other hand, is hidden, regardless of how genuine the conviction in its merits. It just doesn't "look good".

Appearances of fatalism are often avoided by an overt appeal to a philosophy of economic optimizations. To the extent that risk can be operationalized in the form of expected values, the firm operates with respect to risk as it does with any other expense: It optimizes. This is fine for statistical risk, but it does not work for catastrophe risk. Once again, appeals to expected value decision-making in the area of high-stakes risks just represents a form of disguised fatalism.

If some sort of fatalism is chosen, it must be made consistent with the protection of all involved, and must to some degree achieve their at least tacit consent. If we choose fatalism, we must not hide behind inapplicable standards of cost/benefit or other attempts that have the effect of dismissing precautionary dilemmas. While a fatalistic philosophy may ultimately have its merits, it is often the most difficult to own up to via an explicit philosophy of risk. Doing so would at least shine a light on the motives involved. If they are honest ones, fine. If not, they need to be reevaluated in terms of the businesses commitment to protecting the community it serves.

4.7 Post-fact risk management and the *status quo*

Most things we do in life, individually, as a business and as a society, entail benefits, and have countervailing risks. In Figure 4.3 we identify the distribution of probabilities associated with a business venture, to say produce a new consumer product. Monetary benefits are expected for the community and the producer, and they could be

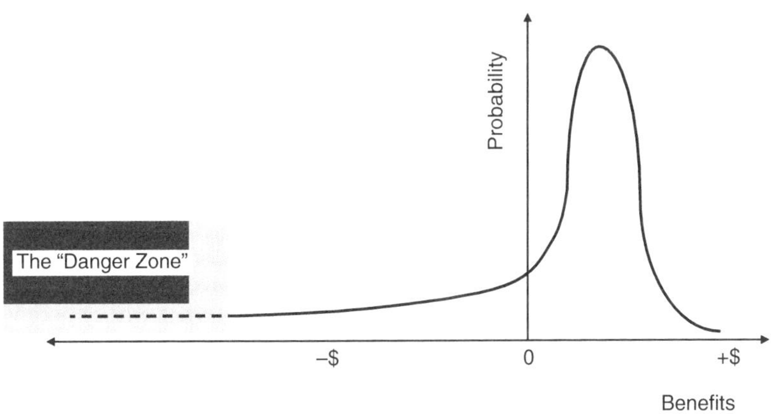

Figure 4.3 Benefits and Risk Potentials: What are the Alternatives?

quite substantial. Against those benefits, certain risks will inure. For example, the mere act of production will require plant and equipment, subject to loss by natural perils. Many of these risks can be treated statistically, and their "average" costs become part of the economic equation for profit maximization for the firm. We judge in this way what statistical risks make sense. We show losses in this diagram as "negative" benefits. And indeed in this view, successful business is about rationally "taking risk".

In some cases, our activities will also entail the small probability of catastrophic effects, both to the business enterprise and the community. Due to the finality and irreversibility of catastrophe, these exposures can not be treated statistically. As we have pointed out early on, there exists no way to balance benefits against the enormous costs of catastrophe. When realistic catastrophic potentials exist, precaution tells us to avoid them, or eliminate them, regardless of potential profitability of the associated activity. In terms of Figure 4.3, we see the leftmost "tail" of the probability distribution of "negative benefits" goes into the catastrophe region, indicating some (imperfectly known) probability of catastrophe. When the possibility of catastrophe exists with sufficient credibility, we avoid the activity altogether. This portion of the figure corresponds to the "precautionary region" (compare this to Figure 3.1 above, keeping in mind that it measures "positive losses", i.e., its scale is reversed).

Where the direct, or opportunity cost (in terms of forgone benefits) of precaution is high, we face the precautionary dilemma (doomed if we do, doomed if we don't). For example, Figure 4.3 might represent the distribution of benefits of the pesticide DDT, the positive benefits including its ability to stop the spread of mosquito borne malaria and other diseases. We have shown how opportunity costs, i.e., forgone benefits, can enter into the precautionary decision process in Section 2.5.

Alternatives assessment attempts to get us out of these dilemmas, hopefully, even before they occur. This allows us to maintain beneficial and stable human progress, without sacrificing safety. In terms of Figure 4.3, alternatives assessment attempts to identify those options that while equal, or approximately equal, in positive benefits do not entail credible catastrophic potentials. That is, activities whose potential negative results do not extend into the precautionary region, or "danger zone".

Note that the benefits in Figure 4.3 usually entail profits, or potential net monetary rewards, for those that support such activities, specifically, the business enterprise. While these rewards are meant to provide reasonable economic incentives to production and innovation, they often become coveted for purely pecuniary reasons. Where these rewards exist, they might not be so easy to give up, especially when faced with the mere prospect for negatives. As a result, attempts to cast the analysis of catastrophic losses into the identify-assess-treat (I-A-T) model, are often simply a matter of attempting to justify profitability, and the *status quo*. The catastrophe level is treated as simply another gradation of statistical risk. The assessment phase of the analysis can not function properly in the catastrophe mode, as probabilities are both very imperfectly known and, due to the catastrophe problem ("in the long-run, there is no long-run"), irrelevant. Risk assessments under these conditions are, as we have suggested above, subject to considerable manipulation in favor of self-interest. Treatment options, which include no treatment, or acceptance of risk, can in this way be tailored to fit the appetite of those creating the risks. A disguised fatalism results, but it is not benign. Whether or not the underlying philosophy of fatalism in this regard is logically justified (say, by the existence of precautionary dilemmas that can not be resolved), the aim of the exercise is to deceive, and profit from this deception. The distinction between the theory of I-A-T risk management, and its reality when applied to catastrophic potential is shown in Figure 4.4.

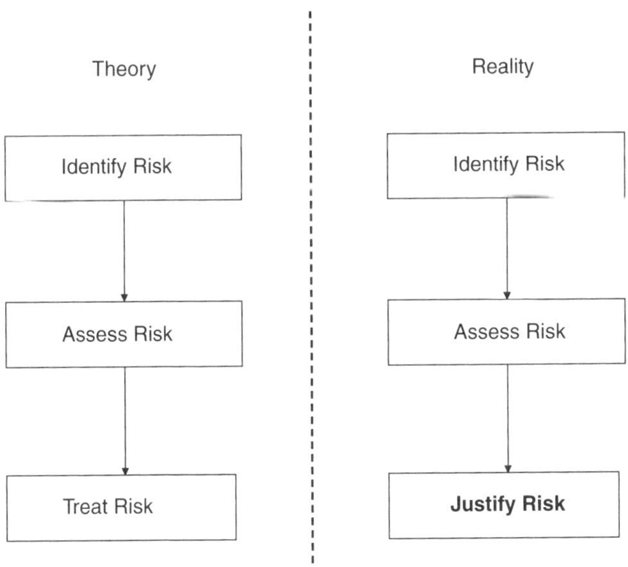

Figure 4.4 Risk Management Theory vs. Reality

We will argue further on that precaution is not inconsistent with progress, and hence profitability. To achieve this progress reasonably, however, requires us to abandon statistical risk management, and its associated I-A-T model, in favor of alternatives assessment. Progressive businesses, and society in general, will realize this.

Businesses that recognize this should formalize their position. An explicitly precautionary philosophy of risk for the business or organization should include a concise statement of the precautionary approach, including a commitment to protection based on the minimax, as well as the conviction to adequately explore and make plain all uncertainties in the process. It should also include a commitment to explore alternatives. In this way, future precautionary dilemmas can be avoided.

Adoption of a precautionary philosophy with respect to high-stakes risk, along with all its trappings, becomes part of a larger corporate vision. As such, things like a precautionary view of scientific progress and development, along with a commitment to alternatives assessment and the avoidance of precautionary dilemmas, has operational implications. In essence, it affects how we "do business". True enterprise risk management, that recognizes as well its responsibility to the community, is thus based on this wider risk philosophy.

4.8 The community commitment

An explicit philosophy of risk must also be articulated at a wider, community level. Democratic decisions depend on open and honest discussion of all facets of society. These include technological progress, and the risks it brings with it. Likewise, basic features of societal governance like land use decisions, require a wider philosophy of the risks undertaken. These wider social values are often reflected in our regulatory and legislative structures, but they go beyond that, ultimately, to the way we lead our lives.

The community's philosophy of risk must be crafted with its citizens, the affected parties, in mind. It should be included in its charter, or mission statement, and perhaps even in its constitution. This reflects the commitment of a democratic collection of individuals that constitutes that community. After all, freedom from catastrophic potentials would seem in itself to be a basic "human right". Addressed in such policy is the community's stance toward public safety and services for its citizens. Those services that are contracted from within (or outside) the community, should be subject to similar risk standards. For example, the community's plan of public safety from disease-carrying insects might contain requirements with respect to the precautionary treatment of any chemical pesticides used for that purpose. On a wider scale, it might address a society's approach to immunization from communicable disease, or its citizens' rights to clean air and water.

Effective precautionary policy includes the requirements that alternatives be addressed throughout the planning process. Beyond commitment, an alternatives assessment strategy needs to be developed. The first step being the identification of situations in which an alternatives assessment is required. These follow from the catastrophic risk assessment process. Do credible catastrophic risk potentials exist? If so, how do we most effectively implement a precautionary approach? Here is where alternatives assessment enters. Our initial tries may result in precautionary dilemmas. This means the alternatives assessment process must rely on the thorough resourcefulness of its participants. The community must work together to encourage the thought process is thorough and adequate. Failure may necessitate an uncomfortable retreat to fatalism. The community must also encourage the participation of all those affected. The assessment process supports precautionary action, which in turn recognizes our community's rights to freedom from unnecessary catastrophic threats. We need to make available both the forum for the assessment of alternatives, and communicate its availability. Business interests will often be an integral part of this process, especially when it is they that are promoting some activity, action or product that may have serious adverse consequences on the community. How do we know we have adequately assessed all alternatives? This question must always remain somewhat "open ended". We never know, for sure. We can only assure that the process is reasonably complete, to the satisfaction of all involved. The same goes for assessing the range of possible effects of all alternatives. Ultimately, the community must be responsible for holding all involved in the process accountable for the results. This means not only those who might promote a possible high-stakes activity, but also all those affected who might contribute to the decision process. Precaution does not entail avoidance for avoidance sake, nor should it serve as a vehicle for those that are opposed to an activity for their own interests, risk potential aside. A certain "fairness" must be assured in the alternatives assessment process.

Last but not least, the successes we achieve through alternatives assessment in achieving precautionary goals should serve as an impetus to further alternative assessments. These assessments would be used in place of inapplicable and possibly misleading statistical risk

1. Identify situations in which alternatives assessment is necessary.
2. Identify and coordinate appropriate intellectual resources for the assessment.
3. Encourage participation of all those affected.
4. Assure [all] alternatives are adequately considered.
5. Hold all participants responsible for the results.
6. Work to make alternatives assessment the "standard" approach to supporting high-stake risk management.

Figure 4.5 Promoting the Community Commitment to Precautionary Alternatives Assessment (adapted from M. O'Brien, "Making Better Environmental Decisions: An Alternative to Risk Assessment", MIT Press, 2000)

assessments based on expected values and the like, as well as an alternative to an "automatic" fatalism in the face of tough risk challenges. The community's commitment to alternatives assessment is shown in Figure 4.5. These basic components of the alternatives assessment process can also be used by individual entities within the community to guide their own alternatives assessment efforts.

Throughout the alternatives assessment process, the community must be aware of forces that may attempt to hinder alternatives assessments. These include special interests that may favor the *status quo* over effective risk management, or those that simply don't know any better, choosing instead to rely on inappropriate statistical assessments. The latter can be counteracted through education in the ways of high-stakes decision-making. The former need to be dealt with via a community commitment to honesty, with strict enforcement of burden of proof requirements on those that would suggest that their activities should proceed unchallenged. Overcoming these barriers depends on our community commitment to the principles of precaution, as well as the commitment of interested parties on all sides to participate in open and honest discussion of high-stakes risk potentials.

There are many examples of communities across the globe recognizing the need for precautionary assessment of alternatives. One simple example is the actions many communities are taking with regard to preserving their collective water resources. A "natural" threat to the quality of standing bodies of water, such as lakes and ponds, is the unregulated growth of invasive plant species, or "weeds". Concern over weed invasions to water resources is often a matter of esthetic concern, or the usability of such waterways for recreational purposes. Invasive weeds look ugly, and they can adversely affect the quality of swimming, boating and other recreational activities. The problem goes beyond esthetics and entertainment, however. Invasive plant species can significantly alter the ecology of the water body, affecting its wildlife, and indeed jeopardizing its future as a viable ecosystem. For these reasons, the spread of invasive plant growths may be considered as an exposure with ecologically catastrophic consequences.

The plight of Rogers Lake in Old Lyme, Connecticut was a case in point. Invasive, non-indigenous plant species, primarily the prolific aquatic plant milfoil, had been inadvertently introduced from other countries, to other areas of the United States, and subsequently to the Northeast region. Its rapid growth in recent years has been sufficient to cause concern over the ecology of the lake, located only a few miles from the Connecticut coastline, in a very ecologically diverse area. Assessing the risk management options, and making the decision, fell on the Rogers Lake Authority, the voting body of which fell under the local government of the Town of Old Lyme. Several community meetings were held, in conjunction with the State of Connecticut's Department of Environmental protection. The framework was one of alternatives assessment, not risk assessment (as suggested by the I-A-T model). The high-stakes potentials involved (including the possible use of carcinogenic chemicals) suggested this was not an issue that could be dealt with statistically.

Several alternatives exist for maintaining the aquatic health of the lake environment. They included the use of chemical herbicides, as well as several mechanical methods. Mechanical methods included harvesting, or culling, the weeds by machine, and dredging. Dredging is a very effective, but very expensive, alternative. It entailed clear precautionary dilemmas with respect to invasive plant reduction and elimination. The cheapest alternative was the application of chemical herbicides. Various aquatic herbicides exist, with the effectiveness at eliminating weeds inversely proportional to their potential toxicity to humans and the natural environment. Harvesting was a middle-of-the road approach, both in terms of cost and effectiveness. Harvesting, for example, does not prevent the reoccurrence of weed growth in the future. While there are chemical treatments that can effectively eradicate the weed, by attacking the root system, these proved to be the most potentially hazardous to humans and the environment. The most benign of the chemical treatments, was no better at preventing future growth than harvesting, though it was slightly cheaper. At community hearings, experts in aquatic environments and the methods of aquatic weed control completely enumerated the pros and cons of various alternatives. The emphasis was on potentials, and no quantitative risk assessments were ever presented. The communities' ultimate decision was not related to the cost/benefit framework, but rather proceeded in an overtly precautionary fashion: The focus was on alternatives. In the end the Rogers Lake Authority unanimously voted to use harvesting.

Of course, other, more complex and far reaching dilemmas exist. Consider the use of nuclear power generation. It provides an alternative to power generation using fossil fuels, which have their own serious negative implications for environmental quality, as well as, for many nations' implications for national security. Nuclear power carries with it negative potentials, some of them enormous. A prudent course of alternatives assessment would suggest that the complete array of alternatives be considered, including conservation, and alternative sources of energy, such as solar and hydropower. Once again, the genuine application of alternatives, along with the recognition of possible catastrophic effects entailed in preserving the *status quo*, or some alternative, based solely on short-term benefits/profitability, may have an effect on how we define progress. High-stakes risk assessment takes us beyond the merely "technical", and into the "philosophical" aspects of risk.

A meaningful philosophy of risk, whether it pertains to the individual, business firm, or community, must consider high-stakes criteria, not just statistical ones. Much tougher issues surface, including matters of ideology that are more conveniently de-emphasized. In some cases, it is strictly a matter of education on the principles and formalization of what we have called the "ABC's", or basics, of high-stakes decision-making. In any case, high-stakes decisions and their explicit criteria cannot be ignored, and an entity cannot have a cogent approach to risk without them.

Underlying our philosophy of risk is our conviction that this philosophy provides for the greatest good. On the wider public level, the requirements for adequate treatment of high-stakes risk may go beyond policy statements and ideological precepts. These requirements may themselves take on the force of law, or at least, be subjected to strong sanctions that promote the adopted risk philosophy. In this regard, we are seeing a rise of explicitly precautionary regulations in the face of an increasingly complex risk environment.

5
Public Policy and the Rise of Precautionary Regulation

When risks affect society at large, our governing bodies have a legal mandate to regulate or otherwise oversee those activities. Risk regulation is part of our government's promoting the "greater good". As such, regulation becomes a reflection of a *society's* philosophy of risk. As we have shown, on this societal level, risks of sufficiently large proportions can overwhelm the ability of us, and our governments, to manage them statistically. As the catastrophic risks at the societal level become all that much greater in the number of people they may affect, the decision process must obviously get more complicated. As recognition of the failure of expected value cost/benefit decisions in the realm of low probability/high-stakes losses becomes more widespread, so has opposition to regulation based solely on expected value. As an alternative, many governments are turning to explicitly precautionary regulation in a variety of areas. While legislation does not necessarily make an idea "right", the risk manager must be aware of the status of legislation in this arena, and the proper place of precaution in it. These rules and regulations will undoubtedly have an effect on how he or she performs his or her task. With a better understanding of high-stakes decision-making criteria, the risk manager is in the position to influence such regulation as well.

5.1 The status of precautionary regulation

Explicit precautionary regulation is still fairly new. Nonetheless, elements of precaution in worker safety regulation go back as far as the establishment of the National Bureau of Mines, in the early 1900s. There, the occurrence of several disasters reinforced the view that mining was an inherently dangerous activity, and those that engaged in it needed adequate protection. The Delaney Clause of the US Food, Drug and Cosmetics Act of 1957 established a precautionary ban on animal carcinogens in the human food chain at the time medical science was furthering our knowledge of the insidious and complex nature of cancer-causing chemicals. It is often suggested that the original US Clean Air Act of 1970 was based on an explicitly precautionary stance. Yet, only in recent years has the precautionary approach become more and more an explicit part of governmental regulation. Much of the new impetus has to do with increased risk, or at least the perception of increased risk, and the failure of cost/benefit based on expected value calculations to deal with these risks. Among new, global risks that we face are global warming, and the development and use of genetically modified food crops. These large-scale risks have brought precaution to the fore.

Though becoming increasingly common in the actual regulatory structure of some countries, especially in Europe, most articulations of precaution appear in the context of international treaties, accords and agreements. This is a natural result of our focus on our widest spectrum of catastrophic risks, those that affect the entire world. These large, trans-boundary risks are the most obviously impervious to simple cost/benefit analysis based on expected values. Not only do these risks have a wide reach, they require the wider global, concerted efforts to attempt to manage them. Such problems are best tackled on a global basis, with consensus treaties and agreements, often created under the auspices of world-wide political organizations, such as the United Nations.

The pronouncements of these treaties are important, as they often become the basis for regional law. For example, the Montreal Protocol of 1987 was a global treaty designed to protect the earth's ozone layer, and thereby the safety of its inhabitants. It did so by requiring the phasing out of ozone depleting substances, including halocarbons, and especially chlorofluorocarbons (CFCs). Up to that time various halocarbons were used as aerosol propellants, and refrigerants, in a wide variety of personal and commercial products, worldwide. The treaty soon became part of governmental regulation in many countries, including the United States (as part of the Clean Air Act).

An explicitly precautionary approach to risk is suggested in the 1987 Ministerial Declaration of the Second Conference on the Protection of the North Sea, 1992 Rio Declaration on Environment and Development (signed at the United Nations Conference on Environment and Development), and the 2000 Cartehagena Protocol on Biosafety. In 2000, the

European Commission adopted a communication on the precautionary approach to risk, paving the way for its inclusion in a wider field of regulation within the European Union.

The various statements on precaution embody what has come to be known as the "precautionary principle". With regard to protection of our natural environment, the Bergen Ministerial Declaration on Sustainable Development issued in 1990 with the cooperation of the Economic Commission of Europe, contains the statement:

> In order to achieve sustainable development, policies must be based on the precautionary principle. Environmental measures must anticipate, prevent and attack the causes of environmental degradation. Where there are threats of serious or irreversible damage, lack of full scientific certainty shall not be used as a reason for postponing measures to prevent environmental degradation.

The Bergen Declaration has features common to all articulations of the principle, including:

1. The emphasis on serious ("catastrophic") losses with irreversible consequences (i.e., those that can not be treated "statistically"). There are no "second chances" to get things right.
2. The requirement that we must do anything in our power to avoid catastrophic losses (the "minimax" principle).
3. Uncertainty about catastrophic loss potentials must be considered in the precautionary decision process.

This last point suggests that we act with precaution when a reasonable possibility of catastrophe exists. This absolutely critical feature of precautionary action recognizes inherent knowledge imperfections in our analysis. The further implications of the existence of such uncertainty is that decisions must never be delayed until we achieve "certainty" as to the distinct possibility or impossibility of an impact. In simple terms: When in doubt, use precaution.

While articulations of the principle vary, the message is clear: Avoid (credible) catastrophic risks. Consider once again the case of global warming. We can easily project catastrophic circumstance from a general, widespread warming of the earth's atmosphere. As for the probability of such an event, there is not sufficient knowledge to permit us to exclude either possibility or impossibility of catastrophic environmental and human effects. Application of the precautionary principle, as suggested, for example, in the Kyoto Protocol to the United Nations Framework Convention on Climate Change (1997), would require immediate precautionary action, in the form of planned reduction of various "greenhouse gases" emitted during the process of industrial production. By waiting until we know "for sure", it may be too late. The dilemma of precaution is, once again, that the costs of avoidance will be high.

Not surprisingly, the Kyoto Protocol, and by association, the precautionary principle, has come under considerable scrutiny, and indeed, criticism. The same arguments against precaution that we have reviewed above have been applied by opponents of precautionary regulation of risk, often quite vehemently. The application of precaution has not only long-run implications, but some significant short ones as well. If your livelihood, or that of your business, or perhaps country, depends on an activity, the suggestion that that activity be curtailed or eliminated may not be a welcome one. Kyoto makes an interesting contrast to the Montreal Protocol, proposed and ratified by most nations of the world a decade earlier. In the case of halocarbons and CFCs, while the loss potentials were clearly catastrophic, a far greater range of "workable" precautionary options presented themselves. The availability of low cost/no cost precaution included the elimination of CFCs as an aerosol propellant. The opportunity costs of such an elimination were relatively low, as most aerosol applications were for the care of personal appearance. There was also the increasing availability of reasonable alternatives, in the form of chemicals that could serve similar purposes yet carried a much higher confidence in doing no harm to people or the environment. What is a matter of so much concern in Kyoto is perhaps not the principle of precaution itself, but the dilemma of what to do about high-stakes risks when the alternatives are just as onerous. Alternatives to fossil fuels remains a distinct possibility, but one with tremendous implications for the way we do business, and run our lives.

As we have tried to emphasize throughout this discussion, the use of precaution is not limited to global, wide-scale risks. Catastrophic potentials at all levels, including the personal and organizational, can be treated on a precautionary basis. Increased attention via effort of regulation and other public policy pronouncements will undoubtedly have a "trickle down" effect to all levels. Today, many businesses in the United States already manage risk with an eye toward implicitly precautionary regulations, such as the Clean Air Act, the Food, Drug and Cosmetics act, and various legislation impacting worker safety. The lesson for all decision-makers from the increased attention precaution is getting from regulators and others concerned with wide scale, world wide impacts is that precaution is a genuine risk management option that deserves serious consideration and study.

Risk managers at all levels can, and should, also learn from the debates that surround potential public policy and regulatory issues that surround precaution. The mechanics of the process, the issues and the debates are all made clearer upon application to "real-world" situations. While global warming, the use of genetically modified crops and the preservation of our environment are all issues on a grand scale, the issues we face in applying precaution are not that much different in the ones the individual business organization faces when deciding on a major retrofit of fire prevention equipment, the treatment of the byproducts of production (potential pollutants), or the decision to launch a new, potentially hazardous, product into the stream of commerce. Likewise community issues such as land use decisions and the implementation of public-safety initiatives, such as anti-terrorism procedures, carry with them the distinct elements of situations where precaution may be applicable. It therefore behooves risk managers at all levels to become not only more aware of the mechanics and rationale of precaution, but to also follow its course through the domain of public policy decisions. There is a lot to be learned from it.

5.2 Strict liability for man-made perils

Despite all the debate, the true fatalist remains, at best, indifferent toward precautionary regulation. In fact, he or she may even actively promote *against* any such regulation. After all, spending time and effort on precautionary regulation goes against the edict of minimizing costs in the face of the inevitable. There remains, however, one practical caution with regard to the unfettered application of fatalism to high-risk situations: The fatalist must be on guard against being "taken advantage of" by special interests. Governmental, or civil, actions may still make sense in that they may provide protection against those that might "abuse" the fatalist's position.

The whole notion of fatalism requires a high degree of resolve. The fatalist must be content to simply accept the hand fate deals him or her, no questions asked. From the standpoint of natural phenomenon, the situation is pretty straightforward. If we choose to play golf in a thunderstorm, we have no one to blame if we get hit by lightning. "Nature" is a fairly neutral participant in the whole matter. The situation is complicated when we consider "man made" perils. Very often, there is a disproportionate benefit to those that create the potential risks. The fatalist is put in the position of being disadvantaged, either intentionally or unintentionally, for the sake of special interests.

For example, a company that seeks to engage in a potentially risky production technology might do its best to make the case that their operations are safe with respect to the community. Perhaps some sort of cost/benefit may be applied based on probability estimates. As we have suggested, any such exercises will have highly dubious results in a high-risk environment. We may not know the validity of the cost/benefit with any level of certainty. Self-interest on the part of the company may bias the result, however "good" or "honest" the intentions.

One way to preserve equity in this situation, without mandating the adoption of explicitly precautionary approaches, is the imposition of *strict liability* for damages on the part of those engaging in the potentially risky activity. Strict, or absolute, liability makes the party engaging in the activity responsible for all damages, regardless of blameworthiness or negligence. Maybe it is more appropriate to say that we are not disregarding blame, as much as holding the mere fact that damages resulting from a presumably safe activity are prima facie evidence of negligence. As a test of their resolve in their cost/benefit analysis, the party engaging in the activity must agree to the strict interpretation. Any residual risk that remains cannot be used to subsequently exonerate the activity. This strict interpretation is based on the mere potential for damage, if the activity should indeed prove "unsafe", or cause a loss. Of course in a truly momentous loss, such "assurance" may be of little value. The point is, the entity is "betting their life" on it, essentially putting up something of extreme value to themselves (their existence) in the event of damages. The basic issue is one of equity: *Parties that do not share equally in the rewards of the risk, must not be required to share equally in the burdens, regardless of their "philosophical" stance toward risk.*

The application of strict liability in cases of high-stakes risk may itself be a valuable impetus for precaution. By implying a potential burden if a "do nothing" approach to risk proves wrong, it helps counteract any "special interests" that may unduly promote fatalism as a disguise for hidden agendas. It helps assure that alternatives will be genuinely and vigorously assessed by raising the stakes for those that fail to do so.

In several areas of environmental regulation around the world, elements of strict liability have been imposed. This makes awareness of precautionary risk management even more important. The trend towards strict liability will undoubtedly spread to more and more public and business decisions.

Consider, for example, the manufacture, sale and distribution of genetically modified organisms (GMOs). A GMO might consist of a food crop, whose internal genetic structure has been scientifically modified to produce chemicals that resist pests. The benefits of such genetically engineered crops stem from the fact that they do not require the external application of pesticides. In effect, they "make their own". We save not only the expense of applications, but also the adverse effects of accumulation of these chemicals in the ground, and otherwise beyond their intended area of application. Nonetheless, the modification of the chemical structure of food crops is an area where the mechanisms are not well known. What *is* known is that such modifications offer at least the possibility of producing allergic reactions in humans who consume the crops. To the extent these crops exceed their intended boundaries, i.e., become weeds, the natural protection of control by pests and other natural phenomenon have been tampered with. These unwanted crops may actually crowd out other, desired crops. Any exposures related to widespread damage of human food supplies, or human health directly, are certainly risks with catastrophic potential.

Cost/benefit analysis simply does not fit when the stakes are so high, and the uncertainties so great. The probabilities of harm are certainly not well known. Indeed, with the loss potentials so high, they may not even matter. Precaution based on minimax would suggest we either undertake safeguards (whose effectiveness may themselves be questionable), or perhaps avoid the process altogether. The onus is on those that continue to support the safety of GMOs. On the other hand, a pragmatic fatalism may suggest we have done the best we can and that we go forward with this very productive product, despite the possibility of disaster. If the position of "no harm" is genuinely held, the producer of such GMOs should have no problem accepting strict liability for any problems that might develop in the future. Prudent risk management on the part of the producer would suggest that the potential inherent in strict liability be evaluated with respect to a precautionary approach. Similar approaches to strict liability in high-risk products can, and are, being applied, worldwide. The increased application of strict liability standards must be considered by informed risk managers when making high-stakes decisions.

The adoption of a precautionary approach by those subject to strict liability reduces or eliminates the potential for action under the strict liability doctrine. To the extent that an exposure is avoided, or its risky components neutralized via precaution-based loss prevention activities, the possibility of strict liability is itself eliminated. By satisfying the burden of proof requirements that an exposure is not risky, and hence not subject to precaution, strict liability has also been avoided.

Application of strict liability may not ultimately clinch the argument for precaution over fatalism. Fatalists, however, need to be assured that their position won't be used by others, quite possibly fatalists themselves, to disadvantage them. Where precautionary dilemmas exist, fatalism may be the only option for progress. We must be sure in that case that these dilemmas are genuine. The only way to do so is via a vigorous assessment of alternatives. Whether such alternatives assessment is "forced" by regulation, or comes about voluntarily, it provides at least some degree of comfort in fatalism. Likewise, all possibility thresholds framed in terms of probability require some degree of fatalism, as the occurrence of *any* event can never be completely excluded. In accepting an exposure as "safe" under these criteria, we need to know the burden of proof is achieved in an honest manner. These assurances become the *sine qua non* of a pragmatic fatalism in the face of risks that we genuinely cannot do anything about.

5.3 Precautionary regulation and free enterprise

We have argued above that the risk philosophy of the business enterprise must be made explicit, at least to the extent it affects a wider public. The perception of the entity may be that risk is an adjunct to progress, and perhaps an inevitable one at that. On the other hand, the general public's risk philosophy may be shaped by the idea that those that would create risk, individuals, businesses, and the other entities that comprise society, may be adequately guided by, even "self-regulated", by economic optimization principles. The same self-guiding principles of optimization that assure the greatest good in the "free-market" should work perfectly well for risk as well, shouldn't they? After all, isn't risk simply a part of production costs, and therefore a factor considered in the overall process of competition is a free market place? The issue is not whether a free market economy is the "right", or most fair, economic system. *Catastrophic risk cannot be easily internalized.* As a result, significant external costs may be associated with catastrophic loss potentials that are simply not part of the normal accounting process of the modern business. This means that, left to its own devices, modern business will always undervalue the costs of catastrophe.

We have seen that fatalism is a "minimum cost" position with respect to catastrophic risk, at least in an immediate sense. It may not be the most prudent approach, but it is the cheapest. This is not to say that modern enterprise is driven to fatalism simply to "cut cost". However, once fatalism is held as a reasonable position, for whatever reasons, the increased costs for precaution will be a hard sell without an obvious economic return to be recognized as a result of these actions. What is the "return on investment" of precaution? Precautionary costs become pure cost, with no beneficial returns reflected by standard accounting practice.

We cannot therefore rely on free markets to solve the issues of high-stakes risks. While not universally accepted, this idea is fairly well entrenched in the thinking of societies, and results in the need for governmental regulation of possibly risky activities. Risk regulation based on statistical analysis of societal aggregates often makes sense. We might, for example, decide to spend public monies on guardrails for public roads, given their statistically proven performance in reducing the severity of the results of on-road accidents. The requirement that auto manufacturers equip their cars with seatbelts (and that individuals use them) is another example of legislation of risky activities on a statistical basis. Pollution costs can be internalized by requiring polluters to pay fines or special taxes. The difficulties enter when catastrophic potentials on a societal level become difficult or impossible to manage statistically, engendering the "catastrophe problem". Loss data is insufficient to apply expected-value appropriately, and the loss potential simply too large to expect any reasonable "averaging" of results over time. Regulation based on statistical optimization of results does not work here. Like the individual, or the business enterprise, society's philosophy of risk must entail a wider view of risk and its treatment as well. This wider view is starting to be reflected in regulatory decisions. The statistical approach is subject to manipulation in the high-stakes arena, and its application simply not logical. A greater and greater part of society is beginning to recognize this. Specialized decision criteria are needed, beyond economic optimization of expected values, to deal with catastrophic risk. Where the wider interest is at stake, these specialized decision criteria need to become part of the regulatory process.

Of course, to the extent that the business enterprise works to voluntarily internalize the costs of risk potentials, outside regulation is "less needed", and hence more efficient. Once again, such self-regulation requires an explicit philosophy of risk. Where that philosophy matches that of the general population, coincident goals make for a more efficient private/public interface. To the extent those interests diverge, for whatever reason, things get complicated. We don't want our regulatory bodies to serve as constant watchdogs of private behavior. And to some extent, things like the imposition of a regulatory climate of strict liability can help make such considerations more automatic. One thing is for sure, to adequately implement a cogent risk philosophy, we must have one in mind. And it is better to develop any such philosophy voluntarily, rather than being forced to do so by the threat of regulation, or other instruments of public policy.

6 Science and Precaution

For a variety of reasons, none of them valid, precaution is viewed as "anti-science", or somehow "backward" in its approach. It is claimed that a precautionary approach ignores science, or at least places unreasonable restrictions on it. Criticisms such as these impede regulation of risk based on precautionary principles, and inhibit voluntary acceptance of precaution. Precaution, however, does not interfere with objective science. In fact, precaution enhances the value of science to a productive and progressive world. What precaution does is make science more accountable when it comes to activities that may expose us to catastrophe. After all, when it comes to catastrophic potentials, even science must be held to a higher standard.

6.1 More science, not less

Precautionary science must first and foremost identify the potential for its discoveries and innovations to cause irreversible harm. Harm is not a by-product of innovation, to be dealt with as an afterthought. The responsible scientist must uphold the same oath as a medical doctor: First, to do no harm. The question the scientist needs to ask is, are catastrophic effects possible (with "possibility" as defined above)? If so, we must refrain from the activity, or seek safer alternatives.

It is sometimes argued, quite erroneously, that the restrictions of a safer science can have negative impacts on the process itself. Precaution can in this way backfire, by crippling the only way we have to get us out of the catastrophic potentials we face, natural or otherwise. Precaution in this way makes ineffectual the very tools we need to defeat catastrophe. We will address this criticism within the framework of science itself.

Science proceeds on the basis of hypothesis. A hypothesis being an educated guess, or a postulate of system behavior, that may ultimately prove right, or wrong. When testing hypothesis, scientist often proceed on the basis of the null, or "no effect", hypothesis (H_0). With respect to the assessment and management of risk, let's say our null hypothesis is, "there is *no* catastrophic loss potential associated with activity x".

The alleged tension between precaution and scientific progress is often framed with respect to the two errors that can be made with respect to potential hypothesis. Following the theory of statistical hypothesis evaluation, these errors may be classified as type I ("type 1") and type II ("type 2"). A type I error occurs when we reject a hypothesis that is in fact true. By rejecting the hypothesis that activity x can't lead to catastrophe, and therefore, say, avoiding it, it is implied that we forego the benefits that that scientific innovation would have brought us. A type II error occurs when we accept a hypothesis that turns out to be untrue. In our case, activity x would be accepted, even though it in fact has catastrophic loss potential. The analysis is summarized in Figure 6.1.

At any given knowledge level, i.e. sample size, there is a tradeoff between type I and type II errors. That is, we can only make one smaller by making the other larger. Progress, it is sometimes argued, requires a bias toward minimizing type I error. Overemphasis on type II error, on the other hand, stifles progress, by rejecting potentially beneficial technologies before they can prove themselves: We can't make progress by being overcautious. These views themselves show a bias toward progress, at the expense of careful, i.e., genuinely scientific, deliberation. Precaution is seen as an "unreasonable" bias toward type II error. We

Hypothesis (H_0): "Activity x has no catastrophic effects".

Type I error: Reject H_0 when H_0 is true.

Type II error: Accept H_0 when H_0 is false.

Figure 6.1 Hypothesis Testing: Type I Error vs. Type II Error

put strong restrictions on science by requiring that everything be proven, beyond a reasonable doubt.

Let's examine the premise that science can only exist if it "takes risk" (in the form of type II errors). First of all, the existence of a trade-off between the two errors assumes our knowledge level is held constant. *We can reduce both types of errors with more knowledge (i.e., a larger sample size).* The goal of science is more knowledge, so that *both* types of errors may be reduced. In terms of uncertainty-modified expression of the minimax, this suggests that our goal is to reduce the interval of uncertainty that surrounds our estimates of possibility of catastrophic loss potentials. When science can precisely identify something as "not risky", given reasonable thresholds for the possibility of risk, we proceed with the comfort of knowing that progress is being made *safely*. At any given state of (imperfect) knowledge, precaution emphasizes type II error. It does so because, unlike the straightforward application of error analysis to statistics, when dealing with catastrophes, there is a possibility the future may cease to exist (the "catastrophe problem"). We therefore have no opportunity to correct type I errors in the future, as is the case with statistical sampling. In sampling, as in many cases of science, a hypothesis can be held until proven wrong. In high-stakes risk management, a wrong hypothesis leads to ruin, and forever precludes our ability to "right" it. When we are uncertain about the effects of an exposure, i.e., when the interval of uncertainty extends into the zone of potential catastrophic risk, we need to act with caution, and work to improve our knowledge so that we can more precisely define the exposure as risky, or not.

A permissive attitude towards type II errors, thereby ignoring precaution, makes science easier. Scientific inventions get to market faster. The scientists themselves gain more accolades (and funding). If we measure progress in sheer number of "innovations", then progress is indeed accelerated. But it is accelerated at the cost of future peril. The tradeoff has to be recognized not as an inherent trait of science, but rather a trait of science that values the number of innovations (and their profitability) over their safety. It is a picture of science that is distorted to fit special, and specialized, interests, not the welfare of the general public. Science in this regard is based on authority: The "say so" of learned individuals. This authority, like any other, can be manipulated. When assessing the true value of scientific activities, we have to ask ourselves if the goal is reducing error, period, or is the goal biased toward making innovations strictly for the sake of making innovations, regardless of their ultimate impacts.

Nothing in the idea of precaution suggests that science be abandoned, or even curtailed. Precaution provides an incentive for innovation, by requiring strict standards with respect to error. We can reduce type I error, but not at the expense of type II errors. To do so, we need to increase our knowledge. For our safety, precautionary restrictions apply until we do. When facing some disease, it makes sense to use science to seek a cure. Until such cure is found, it is reasonable to act in a precautionary manner with respect to that disease. Taking precautions is not an end unto itself. We continue to seek a cure in the mean time. In the same way, exercising precaution does not preclude continuing scientific research. If anything, it accelerates it.

Once we have identified a process as risky, in the catastrophic sense, we also call upon science to determine alternatives. Alternatives assessment for progress requires more science, not less. It is a different view than the one that suggests science be applied to risk assessment only after technological parameters of some innovation are determined. Alternatives assessment is precluded when risk becomes an afterthought. To the extent we must reevaluate our pathways to progress even greater scientific potential is required. Going along with the *status quo* is easy.

In this sense, emphasis on type I error diminishes the importance of alternatives assessment. It suggests we deal with alternatives if and when failure of the original hypothesis ("no harm") is proven wrong. By that time, however, it could be too late. Recognizing type II error places the onus on scientific innovators to "prove" safety, with reasonable regard to the restrictions of possibility. The principle is the same as the placing of burden of proof on those that offer possibly risky technologies and other innovations in the market place. This all makes the search for alternatives an imperative, rather than a mere suggestion. Good science, however, steps up to such challenges.

We have shown in Section 2.5, that precaution, on principle, cannot make matters worse. Where side effects occur, due to the emphasis on avoiding type II errors, they can be accounted for by modifying the decision matrix (see Figure 2.5). So, for example, in Section 2.5, by emphasizing the minimization of type II error of accepting DDT as safe, when in fact it is not, we may increase the malaria threat. Obviously, the ultimate goal here is to reduce uncertainty of both types with respect to DDT. This requires further research. Given that we have information given the possibility of harm of DDT, precaution demands avoidance. This is expressed linguistically, in formulations of the precautionary principle, and can be represented more formally using intervals of uncertainty. Should precautionary dilemmas surface, we seek alternatives.

Science must also be applied to the implementation of precaution: How do we effectively implement precautionary measures? We need to choose measures that work. This problem is independent from the decision to apply precaution. While a great deal of thinking about what makes precautionary measures effective takes place in the analysis of alternatives, technical and scientific skill is required in implementation as well. This is when we may face challenges that may not have been accounted for in the planning phase. Installing sprinklers in a factory building is a reasonable precautionary approach to the risk of a catastrophic fire. Effective design and construction of the system makes sure the protective measures do what we want them to do.

When implementation requires widespread restrictions or bans on the activity, science is needed to properly monitor the enforcement of these restrictions, for both intentional and unintentional violations. A permissive attitude toward type I errors downplays the need for such monitoring. To fully implement precautionary measures, we need full understanding of the participants, on both "how" and "why".

Last but not least, science monitors the precautionary system with respect to ongoing risk assessments. How has our level of uncertainty about an exposure changed, and how does that affect future precautionary action? Scientific progress may suggest that exposures that have considerable uncertainly associated with them become more definite in their effects. This results in continued precaution if we can more confidently deem the exposure is risky, or elimination of precaution, if the exposure is subsequently determined as "safe". While the approach is cautious, it does not automatically preclude technological progress based on continued research. Are the precautionary restrictions too strict, or too loose? That is another question for scientific investigation and democratic review by all those affected. Scientific progress can in this way be directly linked to the process of precautionary evaluation.

The refinement of scientific analysis can be measured via the size of the intervals of uncertainty that surround our subsequent estimates of possibility of risk. The size of these intervals is a measure of the uncertainty involved, and this uncertainty in turn depends on our degree of knowledge. To the extent we are able to bring continuing research to bear, uncertainty, and hence our interval of possibilities, decreases. True progress in science is therefore measured in terms of increasing knowledge, thereby reducing the possibilities. It is not

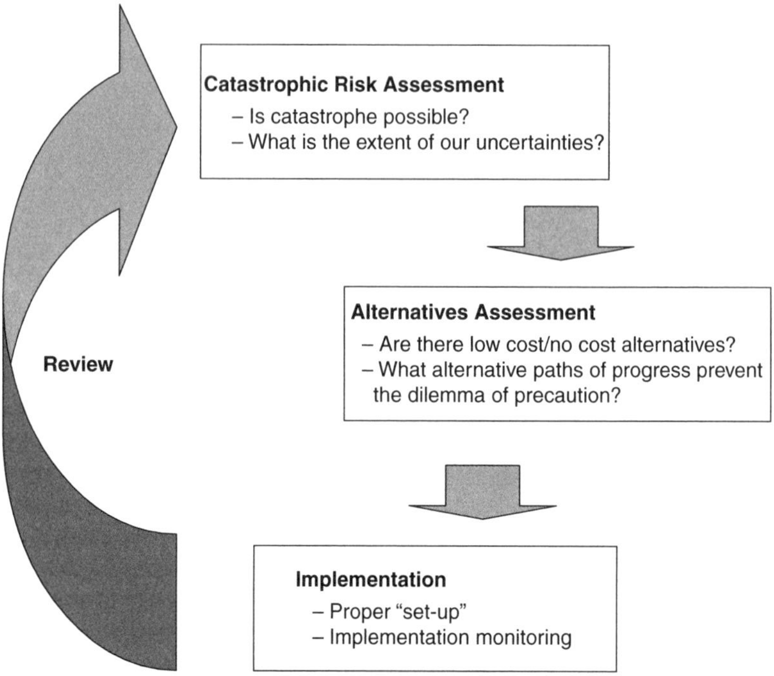

Figure 6.2 The Interface of Science and Precaution

measured by how many commercial successes we achieve, i.e., by innovations or inventions we make. The precautionary process, showing these scientific interfaces, is illustrated in Figure 6.2.

6.2 Probability, decision and the science of risk assessment

Aside from probabilistic considerations of possibility (the impossibility threshold), precaution suggests we ignore probabilities. By doing so, aren't we ignoring valid science? In the late 17th and early 18th centuries, the world began what might be called the *first probabilistic revolution*. Scholars like Fermat, Bernoulli and Pascal pioneered the modern theory of probability. Buoyed by this rise in scholarly interest, the 18th century philosopher and theologian, Bishop John Butler declared that, "probability is the very guide to life." This first probabilistic revolution recognized the basic relevance of probability to life, commerce and science. Tremendous progress in the application of probability theory has been made since then. In modern times, however, we have encountered certain roadblocks to the wider application of probability as the "guide to life" Bishop Butler envisioned. Among them, the recognition that in the real-world, probabilities may be imperfectly known. The second revolution in probability came with the realization of these limitations, and how we may treat these limitations in the practical application of probability theory. We have also come to recognize that the probabilistic treatment of single events, like catastrophes, is problematic. A truly scientific attitude with respect to probabilities requires knowing their limitations. Precaution represents a reasoned response to those limitations. To brand those that reject this bias as "unscientific" is clearly a case of defining science in a way that is most convenient to our interests. Objectivity suffers, and ultimately, so do we.

The science of a formal theory of decision is much newer. Much of modern decision theory has its roots in the post-World War II modernization of the United States. The related field of operations research applied probabilistic and related statistical methods to problems of transportation, distribution and production of goods and services. Decision theory has always had close theoretical ties to the study of economics, and has been used there to describe probabilistic models of consumer behavior ("demand"), as well as costs of production ("supply"). The theory of decision has throughout its history been linked to expected value cost benefit, as presented above, along with extensions to accommodate perceived utility/disutility of monetary gains and losses. So much so that decision theory is virtually synonymous with some form of expected value decision-making. The vast bulk of decision theory has focused on actions that fall into what we have called the statistical realm. In fact, most planning and decision criteria for the modern business firm have time horizons of one to five years, perhaps ten. Most of these decisions therefore fall within the rough bounds of the "10% rule", and can usually be made with some degree of statistical confidence in the results. Decision theory has also been fruitfully applied to the area of controlled statistical experimentation, such as that involving quality control of manufacturing processes. The applicability of statistical decision theory has been vast, and quite effective. This is all the more reason that when precautionists suggest we abandon decision theory based on expected value some wonder whether they are abandoning science all together.

The history of decision theory as it applies to high-stakes events has been rather modest. In fact, precautionary decision criteria in the form of the minimax rule were recognized since the beginning of modern decision theory, say around the late 1940s and early 1950s. Decisions where probabilities were unknown, or irrelevant were referred to as "decisions under uncertainty", as opposed to "decisions under risk", where probabilities are known. The former was regarded as having little applicability, and was usually relegated to a fairly cursory treatment. Duncan Luce and Howard Raiffa's 1953 *Games and Decisions* devote approximately 5 of its 256 pages to minimax decision, declaring it "ultraconservative". In his classic 1970 text, *Decision Theory*, Raiffa likewise gives little space to minimax, believing it unworkable in most situations (due to the prospect of what we have called precautionary dilemmas).

Most of those early decision theory pioneers who dismissed minimax as a workable decision criteria probably had in mind its more extreme version, based on a strict interpretation of "possibility" as any probability greater than strictly "0". As a result, implications such as all buildings in the world should be fitted to withstand the possibility (non-zero probability) of earthquakes does make the strict application seem preposterous. Using a probabilistic limit for "practical impossibility", perhaps modified for uncertainty by introducing an interval valued threshold, we find the principle becomes much more reasonable. In this way, the original theoretic interpretations have been tempered by real-world applications.

Development in the field of formal risk assessment methods roughly parallels that of decision theory. Risky situations have almost always been evaluated using some form of expected value cost benefit. Formal methods for risk assessment that combined engineering and statistics surrounded the rise of modern high-risk processes, including the production of dangerous chemicals and nuclear power. The rise of nuclear power in the 1960s and 1970s particularly stimulated a wave of interest in formalizing methods of risk assessments. Mathematical tools such as event trees and fault trees became the basis of so-called Probabilistic Risk Assessment (PRA). Studies of the safety of nuclear power made extensive use of PRA, with many of the methods translated to other areas of risk. Most of the studies surrounding nuclear power were themselves government sponsored. Out of these studies, ideas of numerical thresholds for acceptability of risk were developed. PRA and the establishment of probabilistic safety thresholds became part of nuclear power safety regulations in the United States and other countries that had nuclear power generation capabilities.

The entire PRA apparatus has been widely criticized for not including uncertainty due to knowledge imperfections. Virtually all PRA studies resulted in point estimates, or probability distributions. A few dabbled in the use of probability intervals and other methods to express uncertainties beyond randomness. Where uncertainty was introduced, decision criteria often drifted toward overtly precautionary approaches, suggesting the wholesale abandonment of expected value decision when it came to such a risky technology.

Despite these defects, PRA remains a widely practiced form of risk analysis in some circles. It is mostly used in studies that are meant to convince the public about the relative safety of some commercial endeavor (PRA, it seems, rarely meets a risk it doesn't like). The complex mathematical apparatus that surrounds such study at least *appears* commanding. Decisions based on PRA are often linked to expected value cost benefit, or precise probability thresholds. Neither provides a particularly realistic, or effective, view of risk in the real-world. As a result, PRA's are often met with skepticism when used to promote real-world policy options with respect to high-stakes risks.

An offshoot of PRA is its application to a variety of hazard risks under the general rubric of "catastrophe modeling". Like PRA, the results of catastrophe models are almost always in the form of point estimates, or well-defined probability distributions. No type of uncertainty beyond randomness is considered. The primary tool of decision applied to these studies is expected value cost benefit. Both on the basis of knowledge of probabilities and their relevance to high-stakes situations, catastrophe modeling and its associated expected value decision calculus fail the test of logical applicability to catastrophic decisions.

Given these failures, it is unlikely that progress in our understanding, or ability to deal with, high-stakes risk will come from advances in the fields of probability and statistics, decision theory (as traditionally construed) or probabilistic risk assessment and its variants, *unless* they are properly modified to (1) deal with the catastrophe problem, explicitly and (2) take into consideration uncertainty due to knowledge imperfections. Dealing with high-stakes risk is not about getting more precise estimates of probability. It is about understanding catastrophic loss potentials, in terms of possibilities and impacts, and selecting appropriate decision criteria that go beyond the realm of statistical losses and expected values. If we ignore these methods, it is because they are not applicable, not because we favor an "unscientific" approach to risk. To suggest that only these methods bear scientific credibility is wrong. If we attempt to discredit the precautionary approach by saying that only these methods are scientific, we are being misleading (intentionally or not).

6.3 Exploratory modeling

In applying science to the first phase of the precautionary process, catastrophic risk assessment, we cannot rely on a simple model of science based on the consolidation of known facts into precise models. More science means better science, or should we say science that is better suited to the problems at hand. Under uncertainty due to knowledge imperfection, our understanding of, and subsequent ability to manage, risk depends on our ability to enumerate, as best we can, plausible alternative risk scenarios. In complex dynamic environments, each alternative represents a possibility that we must consider in our final decision as to how to deal with that environment.

Attempts to assess the extent of our knowledge imperfections in this fashion is the basis of *exploratory modeling*. It is this paradigm of science that must replace the precise, consolidative one if we are to realistically assess risk. Exploratory modeling recognizes that under uncertainty due to knowledge imperfection there may be a collection, or *ensemble*, of models consistent with the data under study. We do not have sufficient knowledge to declare that one, exact model represents the "true" model. Exploration, therefore, results in the specification of *multiple* plausible scenarios consistent with the data (or perhaps rather, lack thereof). No attempt is made to try to summarize or otherwise combine the models into one single "best" model. Rather, the plurality of models is left for further consideration in the decision process.

This process is distinct from those applied when we believe that the underlying model parameters are random variables that can be precisely specified. In that case, we would turn to statistical methods. While exploration does not exclude stochastic models, it does not limit itself to uncertainties which result from randomness.

Exploration also varies from sensitivity analysis. In sensitivity analysis, as properly defined, the values of the underlying model parameters remain static (i.e., known). The *variables* are changed systematically, and the behavior of the model noted. This assumes knowledge of the model. What is unknown, to the analyst at least, is the behavior of the model "output(s)" under controlled changes of its various "inputs". This behavior is specified in the mathematical specification of the model, which in turn requires the model parameters to be known. As the complexity of the model increases, we may need to determine sensitivity numerically, using computer models and perhaps simulation. Numerical methods that attempt to identify the parameters of complex models are, once again, distinct from the exploratory methods we suggest here in that they assume the system is precisely "knowable" (albeit in some complex analytical form).

The methods of generating exploratory scenarios vary. More often than not, they owe more to the process of discovery than to the analytics of verification and validation. In exploratory modeling we seek not so much to discover the unknown, as to discover *how much* we don't know. Discovery is very much a creative process, not just an analytical one. While sampling and search through prospective domains of exploration can be systematized using a variety of analytical methods, expanding those domains remains very much an *ad hoc* matter. The means are often suited to the challenges at hand. As such, many exploratory techniques are developed as part of the application process itself.

Heretofore, perhaps the widest application of exploratory models as we have defined them here has been their use in planning and long range forecasting, both on an organizational and social level. This scenario-based approach was originally developed based on a need to explicitly recognize the effects of epistemic uncertainty on model building in the planning process. While usually based on narrative scenarios, scenario-based planning often incorporates mathematical models as well (e.g., various population growth models). A large and useful body of techniques for scenario generation and application has emerged from such applications. These include the development of structured methods for the elicitation of expert opinion, and robust simulation modeling techniques.

To aid in the exploration process, computers are often utilized. Exploratory methods facilitated by the use of modern electronic computers include interactive visualizations and multiple simulations. In fact, realizing the full power of exploration requires substantial computational power. With the expanded computational capability that is readily available today, due to the increased accessibility of powerful "desktop" computing environments, the idea of exploration is more attractive now than ever before.

The results of an exploratory analysis will not typically be mathematically rigorous, but rather present an imperfect, but realistic, image. The question is if, or to what extent, the results are useful to the problem at hand. The process must ultimately be directed by the questions we seek answers to. In this way, the search space itself can often be made more tractable. For example, in the analysis of risk, we are concerned only with the results of actions that cause losses to the entity (as opposed to gains). Exploration may be further limited to the extent our decisions about risk depend on thresholds. Does the possibility exist (based on the exploration) that we may *exceed* some critical threshold? Exploration exposes epistemic uncertainty. The degree to which we are able to determine the extent this uncertainty permeates our models is a matter of our own ingenuity.

When making high-risk decisions, the risk manager faces considerable uncertainty about the "true" probability of losses. Rather than ask "which to choose?" in the face of uncertainty, the exploratory modeler asks, "*why* choose?" To the extent that uncertainty can be properly circumscribed, and this information carried forward to the decision phase, we have preserved valuable information that can affect the decision process.

Certain intuitive concepts fall into place under the exploratory approach. For one thing, the divergence of opinion often seen among "knowledgeable" experts is accommodated. This divergence does not suggest any of the experts are wrong, but merely that there are multiple plausible candidates for the "true" distribution. Averaging the results often hides this divergence. We lose valuable information about uncertainty. Under exploratory modeling, methods that embrace the variety of expert's opinions are encouraged. An interval estimate may be composed of various expert inputs, for example.

The exploratory view of science for precaution is not based simply on finding what is normal or typical (the "statistical average"). It is wider in scope. It attempts to expand the potential impacts of our actions. We look for what can go wrong. The exploratory approach itself therefore contains an important element of caution that a narrower view of science may lack. This view of knowledge gathering is intuitive. As individual decision-makers faced with the potential of high-stakes threats, it makes sense to seek out all alternatives, especially the bad ones. Our resulting behavior is not based on an average of these results, but rather the worst ones. Emphasis on bad case scenarios makes sense for individual, business and community decision-making. Incorporating it into our world-view of progress is essential for a properly precautionary approach to risk. The analysis of worst case scenarios gives validity to what under normal science may be considered outliers. In consolidative models, outliers tend to be averaged in with other results, and hence, obscured. Our aim is not to overemphasize worst-case outcomes, just to give them the attention they are due.

Exploratory models are far better suited to a realistic view of our world than precise models. Complexity and dynamics do not permit exact specifications. And few areas are as complex as those that deal with exposure to catastrophic risks. For this reason, exploratory risk assessment becomes a natural conjunct to precautionary treatment of potentially catastrophic exposures.

6.4 Science and objectivity

If we take a realistic look at the critics and criticisms of precaution, we find that they are not all that much about legitimate scientific criticism of precautions as much as they are about precaution's potential to disrupt special interests. These interests might be commercial, political, or even personal. Ideally, science should be objective. That is, it must proceed without special interests in mind, based only on the search for truth. In reality, much of science proceeds at the behest of parties interested in specific goals, including monetary ones. While such interests need not in and of themselves taint the process, they can present undue influence on the process. This is especially true in the case where significant uncertainties exist.

Precautionary science often meets with resistance from commercial interests. Scientists in this camp tend to emphasize, indeed overemphasize, the minimization of type I error: The undue restriction on innovation due to the concern over catastrophic risk. It is not the threat to innovation that raises most concern among these critics; it is the threat to *commercial* innovation. Precaution has applications to decisions outside of the realm of commerce. Precautionary principles can and have been applied to ideas such as international conflict and war. Should bans on nuclear weapons testing proceed on a precautionary basis? Is precaution a justification for war? Is there such a thing as precautionary political strategies that might prevent future wars? These non-commercial aspects of precautionary decision receive relatively little attention. Instead, the focus of resources of many individuals and groups seems to be the effects of precaution with regard to hindering this or that commercial venture.

People respect science, or at least the idea of science. Independent, objective, trustworthy. Science is perceived as promoting the common good. Yet, to a great extent it is not above subversion for special interests, especially commercial ones. Scientists are people too. They have to make a living. The same goes for scientific institutions. They have infrastructure and people to support. They depend on income. Science is perceived as a well-spring of innovation, and innovation supports profit. As the saying goes, build a better mousetrap and the world will beat a path to your door (with dollar bills in hand).

The problem is probably not so much one of subverting science toward new directions of profit at the expense of safety and general welfare, as it is supporting "sunk cost" and entrenched systems of production. The automotive industry is geared (no pun intended) to production of internal combustion engines. Research and development into alternative energy sources is expensive. Changing the infrastructure which has grown so large will be difficult. In turn, the automobiles, and other methods of power production, are based on an original source of cheap, plentiful fuel – hydrocarbons. How do you change an industry that has grown so wealthy and powerful based on control of this "essential" fuel source? In the field of chemical production, much is based on post-World War II developments in petrochemicals. The infrastructure sprang up during the war, to supply the vigorous needs of war, and grew ever since. Turning this infrastructure around to produce more environmentally friendly chemicals, perhaps biologically based, is a tremendous endeavor. Hence, the entrenchment of technologies that make precautionary actions against any of them seem "insurmountable". To justify, to hold on to what we have, science becomes the perfect convincer. We have the money to support the science, let's use it. Given the power of commercial interests to corrupt science, we need to be very vigilant. Science, in this modern day, needs to be a democratic process. Simple ideas like precaution lend themselves to a more democratic process, and increase understanding.

This is not to suggest that all, or even most, science is tainted or corrupt by commercial interests. It does suggest that we need to be on guard against that which may be. Special interest must be considered. Dissenting opinions must not only be tolerated, but also encouraged. The interests of all affected parties must be properly considered. The basics of good science for public policy are shown in Table 6.1. In relying on science in making decisions about public policy, we place on it a high standard for reliability and honesty. The outcomes that are at stake are people's lives and fortunes.

Table 6.1 Science for Public Policy

- Who benefits, and how?
- What reviews are in place to assure objectivity?
- Are knowledge imperfections formally considered?
- What processes are in place to assure that the interests of all affected parties are properly valued?

Business interests, as well, must be sensitive to these concerns in their own policies for scientific development. Precautionary risk management interfaces with scientific development in business in a manner that seeks to promote genuine safety, and not just in a way that makes the product or service "most acceptable" to the community at risk. Doing the later implies the community "needs convincing", rather than "deserves safety". By explicitly following the guidelines for good science, we are able to allow the consuming public to judge for themselves.

6.5 Precaution and commerce

The unfortunate thing about arguments against precaution spawned strictly by the fear of commercial hardships is that, in general, they are wrong. Precaution that promotes science is "good for business". Precaution can present opportunities for safe and prosperous economic growth for those that embrace it. Assessment of catastrophic risk requires significant scientific resources. Innovation engendered by the search for precautionary alternatives creates commercial opportunities as well. As does the implementation and monitoring of precautionary measures.

Needless to say, disaster is *bad* for business. To the extent we have prevented catastrophe through precautionary action we have made the world safer for business as well as the individuals who inhabit it. And to the extent that it takes people to make businesses function, protecting them from harm has long-term benefits to those who depend on these people to get things done. A healthy workforce is a productive one. Recent natural disasters have shown that the effects on commerce are widespread, affecting not only those within the physical reach of the disaster. Further research on the widespread economic effects, the "real" cost of disaster, will help illuminate these points even further.

There is no doubt that precautionary dilemmas present the potential for intolerable costs, that both businesses and individual members of society are expected to bear. Incurring financial disaster, or other unfair risk-risk tradeoffs, for the sake of precaution does not make logical sense. The rational precautionist recognizes this. However, the adoption of a precautionary stance does not necessarily commit the decision-maker to dilemma. Dilemmas are possible, but not necessary. We have suggested some ways out of these dilemmas in the discussions above.

Where fatalistic acceptance of risk is all that remains, it must not be considered a financial boon. If we choose to be fatalistic, we need to do so with our eyes wide open to potential abuses for the sake of special interests. We have already proposed increased application of the doctrine of strict liability for those that choose to ignore precautionary warnings, for whatever reason (including "feasibility"). If precautionary dilemmas are something we must begrudgingly face, pecuniary gain for these activities is something a business will have to accept with some degree of circumspection.

That there may be specific commercial concerns for businesses, industries and even communities that rely on techniques and technologies that are the legitimate target of reasonable precautionary concerns is inescapable. In these cases, we must all realize that effective precautionary loss prevention or avoidance is necessary for the greater good. The issue should not be clouded by attacks on precautionary principles simply because they interfere with individual commercial interests. Hopefully, as the study of high-stakes decisions develops more fully, objections based on selfish interests will be more easily identifiable.

6.6 The developing science of precaution

Development of what might be called the "science of precaution" is fairly recent. It was not until the rise of explicitly precautionary regulation in Europe in the late 1980s that the minimax, and its uncertainty modified versions, in the form of the "precautionary principle", started to get serious attention. These regulations were in turn a direct result of perceived practical failure of traditional decision theory to deal with the problems of catastrophic risk in the modern world. We must view the relative newness of precautionary risk management against a background of roughly two hundred years of general probability thinking, as well as roughly sixty years of modern decision theory based on expected value cost benefit. The rise of precautionary thinking was made possible by advances in the theory of ignorance and uncertainty (as knowledge imperfection), which also has a fairly recent development (perhaps, twenty or thirty years). The importance of these will undoubtedly grow over the next few years, as will their status as genuine science.

An important direction for increased scientific investigation, both on a theoretical and practical level, is the improved analysis and assessment of uncertainty due to knowledge imperfections. Modern science often ignores this type of uncertainty. Precision is valued. Imprecision is something to be "dealt with". In the mean time, it is hidden. The great potential for further scientific study in the areas of catastrophic risk assessment lies in systematically addressing issues of knowledge imperfection. How can uncertainty due to knowledge imperfection be identified? We have identified the use of intervals for specifying this type of uncertainty. How are intervals determined? How are they verified? This is the type of focus modern science needs to take. By formally recognizing knowledge imperfection, many of the anecdotal and common sense features of precaution become amenable to specification and study. In Appendix A, we present a working model of precaution based on the theory of fuzzy sets. Fuzzy sets are a generalization of intervals which allow a richer specification of uncertainty due to knowledge imperfection.

A formal analysis of uncertainty due to knowledge imperfection will also help us specify rough thresholds for the concepts of "catastrophe" and "possibility". These, as we have shown, provide a more accurate view of how precaution must be implemented in an uncertain (partially known) world. We have applied the theory of approximate reasoning under knowledge imperfection, using interval assessments, to suggest a definition of possible risk based on probability. This theory can be applied to other properties of the world. As suggested above, perhaps our definition of possible risk is not based on probability at all, but other some other physical aspects of the world. This is what further research will need to determine.

A great deal of confusion, miscommunication, and indeed resistance, to precaution stems from the fact that precautionary principles have been viewed from the narrow perspective of precision. Applying precision means we utilize exact thresholds for impossibility. It means we measure probability precisely. All these things ruin the applicability of the principle. They present a misinterpretation of precaution that sets it up for criticisms on the basis of "workability".

It is likely that many who champion the precautionary approach to risk are themselves unfamiliar with the formal aspects of modeling uncertainty due to knowledge imperfection. Instead, they often attempt to justify the principle by applying a formal apparatus suited to the precise determination of probability, just as their critics do. As the framework of precision is not applicable, these attempts are doomed to failure. The uncertainty representations entailed in the precautionary principle cannot be formalized using the precise, or "crisp", theory of probability. So while epistemic uncertainty (about consequences and possibilities) is at the very root of the "precautionary approach" to risk, many of those on both sides of the debate have no idea about the formal properties of this type of uncertainty.

That precaution functions perfectly well on an intuitive level is not surprising. We use approximate rules-of-thumb to make decisions all the time. If it is formalization we seek, then we may turn to the well-developed theory of fuzzy sets and its related logic. Such formalizations would be useful, for example, if we wanted to operationalize the principle for the development of computerized *expert systems* for risk management. To understand the treatment of high-stakes risk, we need to understand uncertainty.

Development of a science of precaution is also very much about how we "do" science. The emphasis in the scientific development of precaution is first on the assessment of catastrophic loss potentials, and that entails significant uncertainty beyond the idea of randomness. Armed with this more far-reaching view of uncertainty, we tackle the problem of identifying catastrophic risk exposures. The biggest challenge comes after these risks have been identified. It then becomes a matter of seeking adequate alternatives so as to *avoid precautionary dilemmas*. The landscape of high-stakes risk is dotted with these dilemmas. We either resolve them, or live with them (i.e., become fatalists). It is this development of precautionary alternatives to conquer precautionary dilemmas that presents the toughest challenge to a science of precaution. The task is made all that much harder when we consider that "normal" science often considers the challenges of high-stakes risk as a byproduct of progress, not as a fundamental guide to sustainable progress. This is, once again, a side effect of a view of scientific progress that has to some extent been tainted by commercial success. All of those with a stake in scientific progress, including those who may benefit commercially from such progress, need to realize that catastrophic consequences could bring any such progress to a complete halt. It is therefore in everyone's interest that a science of precaution begets precautionary science.

7
Communicating about Risk

Understanding the risks we face requires communication on many levels. On the technical level, we need to understand the basic concepts of risk, including probability and randomness. Often, these more technical aspects of the concept of risk are obscured by our use of the term in subtly different contexts, and with different meanings implied. An important first step to effective communication about risk is to understand the various meanings of the word "risk". We then need to be specific about how we use these meanings in the context of communication. We also communicate about technical factors that cause, or expose us to risk, including various notions of cause and effect in the risk domain. Because of the complexities involved, such communication must invariably take into account uncertainty due to knowledge imperfections.

Catastrophic risks have the potential of affecting many. We need to be able to communicate about risk to those potentially affected, so that they can become rational participants in the risk management process. Rightfully, if you may be adversely affected by risk, you should be a participant in the process. This means establishing an interaction with those affected, rather than simply providing information (or possibly, misinformation). Interactive communication about high-stakes risk promotes effective risk management.

7.1 The meaning(s) of the word "risk"

Early on, we defined risk in terms of hypothetical draws from an urn, or bowl, filled with colored balls. This simple "thought experiment" defines the most basic components of the generalized notion of risk: Anything entailing the probability of an adverse outcome. We pointed out as well that various combinations of probability and consequences can lead to different varieties, or conceptions, of risk. Colloquially, we often use the word as a simple synonym for danger, or peril: Skydiving is risky. The same word has a less ominous connotation when we warn our friend that by eating spicy food, she risks indigestion. In the first case (skydiving), risk clearly refers to the potential for catastrophe: High-stakes losses that lead to loss of life, or result in a serious impediment of health. In the later case (spicy food), we describe an unpleasantry, or inconvenience. Both involve costs, those involved in skydiving being largely, perhaps immeasurably, greater.

Our first step in effective risk communication is carefully differentiating the meanings of risk. Important differences exist between statistical and catastrophic losses and their treatment. It is important that when communicating about risk we do not obscure those differences. "Avoid risk": This precautionary motto applies well to catastrophic risk, but not to statistical risk. Avoiding all statistical risk will result in measurable inefficiencies. On the other hand, the suggestion that risk be treated statistically makes sense up to the point where these risks entail the "catastrophe problem". Making the distinction sounds simple, but it is not. We often tend toward generalizations, even in highly technical representations of problems about risk. We may announce boldly to the company's board of directors that our future should be guided by a "risk taking" attitude. The meaning of risk here is clearly statistical, in that it urges the company to move forward by taking calculated statistical risks that will demonstrate their rewards over time. If we go in front of the board and urge the company to accept more danger, we would get an altogether different reaction. Our usages of the multiple concepts of risk are subtle, so we must be always aware of how we are using them.

In the financial world, risk is often used synonymously with statistical variability. Chance variations from some intended result represent risk under this interpretation of the word. Measurement of such risks proceeds mathematically, based on mathematically (or experimentally) well-defined statistical distributions of the probability of loss. Variability may only be tangentially related to danger, and often involves no seriously adverse effects at all. The fact that my very small portfolio of stocks entails some modest risk (i.e., variability) may have little or no bearing on my financial health. It may only serve as a guide to my expected returns over a sufficiently long period of investment.

Measures of risk in this regard are limited to the statistical domain. Expected values, as well as statistical variations around this expected value, provide a suitable measure of risk with regard to long-run expectations. As we have pointed out, such measures make no sense in the domain of catastrophic losses. Sheer potential serves as a guidepost to action in catastrophic situations. When we have achieved or exceeded the catastrophic threshold, we take precautions. So while the basic meaning of risk remains the same (a draw from an urn of colored balls), severe consequences dictate specialized actions. We will continue here to focus on high-stakes, catastrophic risk. We must keep in mind, however, that many definitions of "risk" exist when communicating about risk, in general.

7.2 The lay perception of risk

In discussions of the methods and goals of risk communication, a distinction is sometimes made between the layperson's, or general public's, perceptions of risk, and that of the risk "experts". The distinction is often made to help reinforce the need for the education of the lay public on the ideas of risk, the fundamental assumption being that the general perception of risk is somehow lacking. Undervaluing the general public's ability to understand risk is perhaps the biggest mistake one can make in attempting effective risk communication, especially in the high-stakes domain. Professional risk analysts and risk managers have both a technical and intuitive understanding of risk, gained from the experience of directly working with a variety of risks, as well as specialized education and training. For these reasons, they have a better understanding of the formal structures of risk, and how these may (or may not) apply to the real-world. This does not mean that their perceptions of risk are any more accurate than the general public's. The ability to perceive risk, and its potential effects, is very much an intuitive exercise, gained through general experience with our complex world. In studies of the ability to assess the probability of events in this intuitive setting, experts do not necessarily outperform lay persons. And this is understandable. We have shown, in the high-stakes arena, where intuitive perceptions of risk are the norm, there are no technical devices available for assessing risk on any precise basis. This is distinct from the statistical world, where we are able to specify random events with a high degree of theoretical precision. In the statistical world, there is no such thing as "lay perceptions" of risk. We may all experience certain biases and use simple heuristics with respect to assessing statistical risk, some of them accurate, some not. Lay perception does not matter. If we need to find exact solutions to statistical problems, we turn to the well-developed theory of probability and statistics. We do not use perceptions, expert or otherwise, to determine the statistical proportion of defects in a large batch of machined parts. We use statistics.

Our intuitions about risk in the high-stakes domain reflect some extremely complex and dynamic processes. By all accounts, our intuitive judgments of likelihoods and the possibility of risk in high-stakes situations take into consideration knowledge imperfections. Knowledge imperfections are reflected in the use of words such as the "possibility" of risk. Numerical variables are replaced by linguistic variables that convey the inherent uncertainties involved. Our knowledge about linguistic variables and how they function in the world is gained via experimentation and experience. It is not just about counting positive instances, and making pronouncements based on these statistical assessments, however. The type of inductive logic that lies behind this deeper knowledge is more complex. It includes analogical thinking. It also encompasses the idea that there are limitations to our ability to know things precisely, and that we must take these limitations into account.

Linguistic variables allow us to deal with the inherent imprecision of the high-stakes world. In this way, they help us to more accurately represent reasoning about catastrophic risk. Based on what we have called the second revolution in probability (Section 6.2), we are able to formalize linguistic variables into theoretical structures that are just as rigorous as those used in the study of precise probability associated with the analysis of statistical risks. We do so using interval estimates and other structures that preserve the inherent uncertainty due to knowledge imperfections, yet let us analyze the process using formal scientific methods.

As we have suggested in Sections 3.2 and 3.3, human beings face high-stakes risk with an intuitive model of the precautionary region, or "danger zone" in mind. They match this rough, but useful perception to approximate ideas of an events "possibility", in terms of interval probabilities, to come up with an appropriate assessment of risk. Based on these perceptions, precautionary risk management is exercised.

It makes sense that these assessments of risk must ultimately come from those affected. Of course, this does not mean that the general public need become experts in chemical exposures to health, or workplace safety. These are informational inputs into the risk assessment and perceptions process that need to come from the "outside". These inputs, however, are just a component of risk perception. The affected persons may also need guidance on how to translate technical findings into possibilities of adversity and *how* (and *how much*) they might impact them. Yes, this does require a good deal of intellectual effort on the part of concerned stakeholders. This effort is part of dealing with the complex issues of life.

When the public ignores cost/benefit based on expected values, the "experts" often fault the public. The implication is that they need to be educated in the ways of expected value decision-making, and the assessment of probabilities. They just don't know. Studies show that the general public does indeed tend to ignore probabilities of high-consequence events, focusing instead on their impacts. This is sometimes viewed as a sign of ignorance by those risk experts used to dealing with risks in the statistical realm. Ignorance indeed, but ignorance that simply cannot be remedied by technical means. No one can know the probabilities of low probability/high impact events with any precision, due to the complexities and lack to data. When faced with the catastrophe problem people also intuit the defects of arguments that rely on outcome in the long-run. It is not just a matter of educating people in the theory of probability and expected value decision-making. When probability theory is useful, as in the case of losses that occur frequently enough that we can use statistics, we call in the statistical specialists. These same specialists are of no help in the high-stakes domain, as their methods are inapplicable. And in the fields of probability assessment in the intuitive domain, risk experts and the lay public are on par.

Professional risk managers must understand that they can learn from the general public, as well as improving that publics understanding of risk. What factors are important to the people affected by risk, and how do we craft a risk management strategy to fit? This is not to say that people don't make mistakes, or have biases that may negatively affect the process. These are the areas where expert input is truly valuable. Yet, it is valuable only within some reasonable framework for high-stakes decisions. To fault people for not using expected value decision-making in an area were it is clearly inappropriate is wrong. Pushing these methods in a domain where they are clearly inapplicable can only mean that the analyst is completely ignorant of their inadequacies, or that some special agendas underlie the process. With respect to the later possibility, effective risk communication is not about using "expertise" to persuade, or indoctrinate. It is about honesty in the selection of appropriate decision methods, and a respect for the perceptions of those affected by those high-stakes risks under study. As for being ignorant of the special characteristics of high-stakes risk, such a lack of awareness would hardly qualify one as an "expert" in the domain.

7.3 Effective risk communication

Making and understanding arguments about high-stakes risk, whether they are for or against some sort of exposure, or expenditures to manage the risk of such exposures, requires communication about the facts. Informed decisions, decisions that embody the essential element of trust, require all involved to be as up front about the exposures to risk as possible. To this end a specialized field of endeavor has arisen around the need for effective risk communication. It attempts to identify things that promote, as well as obstacles to, effective communication about risk.

While the act of communicating facts seems fairly neutral, risk communication efforts, more often than not, seek to convince, and not merely give facts. Tremendous gains are to be had in the introduction of a new idea or technology. The manner in which we assess and treat any potential catastrophic risks from such exposures is crucially important to those that would realize these gains. As a result, we find tremendous resources expended on attempting to convince society that either the risk is not credible, or that a reasoned fatalism should prevail towards it. And these efforts are not universally focused on proving "no risk". In some cases, financial gains and other issue of self-interest promote the finding of risk, with possible precautionary repercussions. For example, it is sometimes suggested that countries use precautionary legislation based on undue amplification of risk to institute protectionist policies against imports for another country.

A remarkable amount of risk communication has become more a matter of "public relations" than the simple dissemination of scientific information. It has become a battle for the "hearts and minds" of the general public. Expenditures in this area are tremendous, and might even rival those spent on the analysis of risk itself. To a great extent, this is a reflection of modern society, where genuine efforts to provide convincing evidence are replaced more and more with synthetic arguments that focus on emotions, rather than the intellect. Availability and impact of modern media probably have a lot to do with this.

While we have presented a strong case for why precise numerical estimates and expected value calculations do not work, the apparatus that surrounds the "classical" approach to optimization remains incredibly attractive to the human mind. Humans seek order, and the equations of mathematical optimization provide an almost celestial order. The techniques applied are a direct outgrowth of optimization techniques in the physical sciences, where these theories have led to tremendous practical advances. With the availability of computer technology, a new element of attraction to the "technological" solution to our risk problems has emerged. We should not let the allure of technological solutions obscure the fact the catastrophe problem is not amenable to a solution in terms of optimization, computerized or not.

Above all, we must be honest about the process of high-stakes risk assessment, and what it can and cannot do for us. When dealing with high-stakes risk, the statistical approaches we use to deal with the more mundane are not applicable. We need to be clear about this. We must use decision criterion suited to the high-stakes domain. We must also be very clear about the uncertainties involved. Our desires to exercise control over some precisely defined world are strong. These desires must not overcome reality, however. And that reality is that precision does not exist. The uncertainties, due to knowledge imperfection, are large. We need to face them, and incorporate them into the decision process – not ignore them.

Communication about risk is not about rationalizing precautionary dilemmas. We can not dismiss or diminish credible catastrophic potentials based on unduly precise probability estimates, and the appeals to cost/benefit analyses. We have shown that uncertainty due to knowledge imperfections in the process of assessing high-stakes risk can be effectively conveyed using intervals, and their various generalizations. By conveying potential risks in this fashion, we allow both specialists and the general public to determine potential overlap with the precautionary region. Unduly precise point estimates only serve to confuse and possibly confound important decisions in the high-stakes realm.

The scientific process of obtaining reasonable estimates of potentially catastrophic risk, however imprecise, should also be made plain. This involves dissemination of information of the exploratory modeling process and its components (e.g., qualitative event trees that trace logical progression of loss events, but do not specify exact probabilities). Relevant research on causal relations should also be made available, including "dissenting opinions". By presenting all sides, we need not appear ambiguous or wishy-washy in our approach. As noted in the Chapter 6 Science and Precaution, the science behind high-stakes decisions needs to be accurately represented.

These transparent approaches make sure that worst-case scenarios are properly considered, and not diluted away via a reliance on averages and summary measures that do not properly address the potential for catastrophe. When the potential for catastrophe exists, in any situation, at any level, it is not a matter of if we should make this potential apparent, it is how. Worst-case scenarios, exploratory models and interval estimates are all good candidates, and can often present a particularly realistic view of risk when used together as part of the risk communication process.

The factors that influence perceptions of those affected by high-stakes risk potentials must also be understood as part of the communication process. Proper respect should be paid to these factors in that they represent an ideological perspective, not purely an emotional or subjective one. Individual characteristics of the exposure to potentially catastrophic consequences can, for example, affect the acceptability of such exposures. This in turn influences the choice of high-stakes criteria. We will argue in the next chapter that such acceptability criteria are fundamental to understanding the potential for a "mix" of fatalism and precaution in our modern world. We summarize some of the features of effective risk communication in Figure 7.1.

In this discussion of risk communication we have been careful to point out the potential for specious arguments to influence. This is not meant to imply that all differences of opinion are a result of conscious attempts to mislead or manipulate. When uncertainty is great, we can't say that if one side of the debate is wrong, the other knows it. Honest differences of opinion certainly can and do arise. In fact, when uncertainty is great, we would expect such differences to arise with greater frequency. Can we separate honest differences from dishonest ones? It is often difficult to determine how genuine a belief is. While the existence of ulterior motives does not mean that they in fact influence communication, they are a factor that must be considered as at least

- Use decision criteria suited to the high-stakes domain
- Understand and communicate uncertainties
- Understand and respect factors of risk perception
- Get feedback from stakeholders (communication must never be "one way")
- Use communication to inform, not to manipulate

Figure 7.1 Effective Risk Communication

having the potential to do so. In assessing the value of risk communications of any sort, it makes sense to check out the "pedigree", if you will. Do those promoting a particular risk, or risk approach, in a risk communication have any significant monetary gains (or losses) at stake? If scientific expertise is brought to bear, how have the individuals involved been compensated? If a research or scholarly entity is involved, what are its affiliations? We need to be wary as well of the properties of "good" science in this regard, as specified in preceding chapter (as summarized in Table 6.1). Again, we cannot say that anyone who promotes a risk that they have an interest in is doing so blindly. The concept of reasonable burden of proof assumes this burden can be overcome. That's a pretty basic part of being open minded, and in turn, a basic part of progress. But it is likewise improper to take every bit of risk communication on "face value". The stakes in risk debates can be tremendous. Our sense of what constitutes scientific credibility in these cases must be extra keen.

All things considered we need to recognize that some forms of misinformation and manipulation have become a feature of risk communication. Given the importance of the subject, this area of risk management is probably worthy of some more concentrated study. Dealing with risk is tough enough without intentional obfuscation. Understanding the criteria of effective risk communication helps make us more aware of these potentials, and how to avoid them. It also helps make us more honest communicators.

We have emphasized "external" communication here, that which involves the entire affected community, or the public at large. Effective risk communication is an essential part of decisions that are internal, to the individual, the business entity, or those responsible for the health and safety of the community (for example, regulators and our political representatives). As individuals, we need to understand risk to be able to properly face it. Understanding effective risk communication permits a credible dialogue on things that affect us. The business makes many internal decisions based on the risks it faces. To do so, it is critical that management be properly informed of those risks. Effective high-stakes risk communication is an essential part of the enterprise risk management process (ERM), as described above. Each day, executive management of the business enterprise face potential exposures to catastrophic risk. It is like drawing from our bowl of colored balls: Selecting that one ball could mean disaster. It is imperative that these executives have some rough idea of risk associated with each "bowl" they are drawing from (operational, strategic, hazard, financial), and its potential for severe adversity. Doing so allows them to govern effectively, by choosing the best risk management strategy for the conditions.

On the community level, internal communication is a crucial part of the regulatory process of government and other social organizations. Our leaders need to be aware of catastrophic loss potentials so they themselves can help guide us through an appropriate program of high-stakes risk management, and make informed decisions when legislating in this regard. An informed body of government may be society's best hope for a cogent global strategy of high-stakes risk management.

7.4 The feedback process

The value of understanding risk perceptions lies not in their power to manipulate risk decisions in favor of some special interest, but rather as a guide to help us down the path of effective risk evaluation. Risk managers at all levels must factor these perceptions into the process. To do so requires feedback from all stakeholders. So, for example, the corporate risk manager would do well to periodically evaluate how any risks they may present to the general public, and other stakeholders, are perceived. What are the catastrophic potentials? Is catastrophic risk judged as possible, or not? What level of uncertainty is involved? What evidence do the stakeholders require to make effective decisions about risk? Have alternatives been adequately and completely assessed? This gives any entity proposing new ideas, approaches and technologies a feel for how the public will respond, based on rational risk decision criteria that apply to high-stakes risks. These can be matched to internal assessments, to guide risk policy accordingly. The exploratory process of risk assessment can also be more effectively guided with this information. Development of ideas becomes iterative with respect to risk, and in fact, more democratic.

Risk perception surveys at all stages of idea development, implementation and throughout the history of the product or service help guide the risk management effort. Not by suggesting areas were the public need "convincing", but rather in ways that the producer can adjust the product or service, throughout its "life", to better suit high-stakes risk criteria. This in turn goes a long way to satisfy burden of proof requirements under a precautionary approach to risk, whether it be mandated by government regulation, or voluntary.

Risk communication must always be "two way", with feedback from stakeholders being a crucial part of the process. These external assessments can provide important information. Above all, risk communication should honestly represent the underlying risk philosophy of an entity. It should not simply be a "public relations" effort meant to manipulate the views of stakeholders. Catastrophic risk potentials are too critical to be the subject of efforts to influence or cajole, regardless of how well intended those efforts may be. A key factor in risk perception is trust, and this must be reflected in the process of risk communication. An effective feedback process goes a long way to building this trust.

The feedback process can also benefit from monitoring by "outside" parties, such as government regulators and legislators. As these parties are sometimes called upon to intervene in risk decisions, it behooves them to know more about the concerns of all involved, as well as get a balanced view of the exploratory evidence. This understanding is imperative for both the effective establishment and enforcement of precautionary regulation (Chapter 5). Informed monitoring of risk communication by regulatory bodies can also help curb the intentional, or even unintentional, dissemination of misinformation about catastrophic risk potentials. Risk communication often seems to be subject to a sort of "free-speech" doctrine: We can say what we please about risk. Free speech can be abused when those in positions of trust use it to intentionally misinform about high-stakes risks. Rather, the cannon of "truth-in-advertising" should apply: You can't say it if it is not true. With respect to high-stakes risk, truth-in-advertising means that those that promotes potentially risky endeavors must recognize that the burden of proof is on them. When in doubt, the obligation is to err on the side of caution. Truth in risk assessment is, however, a tricky concept. There are those who would argue instead that we should be able to say what we *believe*. As we have seen here, beliefs about high-stakes risks are complicated. When others are potentially affected by those beliefs, we in turn need to make sure those beliefs are shared among all involved. If not, a dialogue among interested parties must ensue, to get to the bottom of things.

There should also exist sanctions against misguiding statements about risk that promote self-interests (on either side of the debate). Misinformation impedes the process of honest risk assessment. Of course, how such actions are to be detected is a complicated matter. Failure to adhere to logical principles of high-stakes decision is certainly a telling sign.

8
The Future of Risk

Perhaps fatalism would not be a bad alternative if the level of risk in the world was constant. If this is the way it has always been, and always will be, there does not seem to be much point in doing something about it. Significant evidence points to the fact that the level of risk in the world may well be increasing. We'll look at some of that evidence here.

High-stakes risk presents challenges. The ability to optimize our response, based on scientific assessment of probabilities and application of the classical economic apparatus, is attractive, but fiction. We can't develop a realistic philosophy of risk, be it personal, corporate or societal, based on statistical analysis and expected value decision-making. We need to focus on logical high-stakes criteria, precaution and fatalism, despite the difficulties in their application. Perhaps effective high-stakes risk management requires some blending of the two, in that different risk characteristics require one or the other approach.

Our ability to shape our future depends on our ability to understand risks that can, in an instant, change that future. Against a natural background, modern humans seek progress. Understanding what risks that progress brings with it, and what to do about them, is a key element to sustaining this progress. Hopefully the discussions here will help make that pathway clearer.

8.1 Is risk increasing?

Before we address the issue of increasing risk, let's look at a more fundamental one: If significant risks are so "unknown", how do we know they even exist? Statistical methods address what is known, and therefore they represent all we need to effectively "manage risk".

Discussions of fatalism and precaution are philosophically interesting, but practically, irrelevant in this case.

We have to recognize that there is a crucial difference between knowing something exists, and being able to precisely delineate it. When the potential consequences can be defined as "high", we need to pay attention to that potential. Sadly, the world periodically presents us with events that are a great detriment to life and property. Those that, nonetheless, steadfastly refuse to recognize the existence of any risks but those that can be statistically "proven" can count themselves as members of the fatalistic (i.e., "do nothing") camp. The modified credo of these fatalists becomes: Significant risk may or may not exist. If it does not exist, there is no problem. If it does, we can't do anything about it anyway. So, why worry?

The growth of risk and the future potentials for increased risk have been a matter of some debate over the years. The validity of the suggestions herein is not dependent on the premise that risk is growing, or that it will continue to grow in the future. The fact that growth in risk exists, however, elevates the importance of these observations. We would argue that such growth has indeed taken place over time. One hundred and fifty years ago, the cannonball was the "weapon of mass destruction". Today the threat of nuclear weapons is global. A favorite statistic of those that suggest the world is indeed a "better place" now than ever before is life expectancy. The statistics are compelling, when one considers the average life expectancy in the United States was 50 at the turn of the century (especially compelling to those over 50). "Quality of life" arguments aside, the question becomes whether that expectancy might reasonably be much higher, were it not for adverse environmental factors introduced for the sake of "progress". While it is easy to see how improved medical imaging has increased life expectancy, it is hard to see how carcinogenic chemicals in dry cleaning and hair dyes have. The suggestion seems to be that all of progress as we now know it must be looked at holistically in its contributions to life expectancy. We can identify those aspects of progress that have arguably prolonged human life spans, and those that might have reasonably shortened them. That the two must go hand in hand is far fetched.

A mere look at the exponential growth of Gross Domestic Product in the United States over the last 50 years would suggest, prima fascia, commensurate risks are expanding (Figure 8.1). Expanding faster than we can control them?

The answer lies in the hidden "bottom" of the iceberg model of risk, and is not, unfortunately, as obvious as some would have us believe.

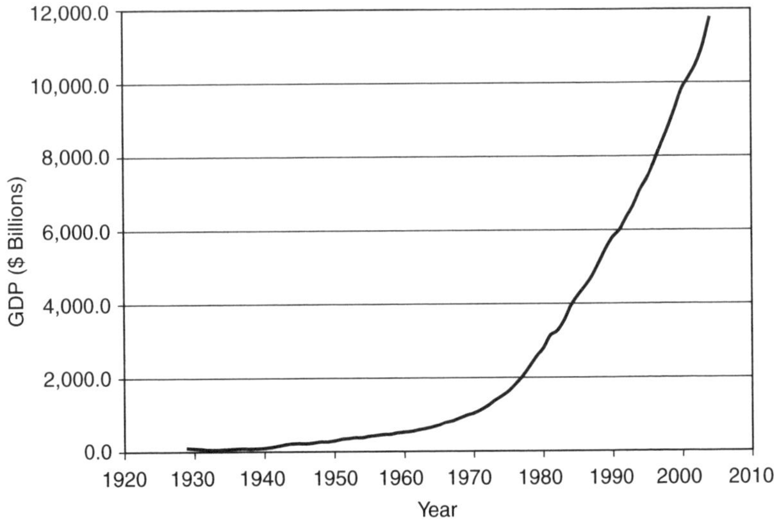

Figure 8.1 Growth in US Gross Domestic Product (GDP), 1929–2004

The expansion of risk lies not only in technological, and other "man made" exposures, but in age-old natural perils as well. Earthquakes, flood and windstorms have been around since the beginning of time. However, unprecedented population growth and industrialization have lead to increasing human utilization of natural disaster prone areas.

The great tsunami disaster of 2004 is a sad case in point. Increased population growth and the need to center economic life around the ocean means that considerable populations around the world are subject to loss via tsunami inundation (this goes for industrialized countries as well). Our technological response to risk management in this realm has been early warning via ocean-based seismic and wave height sensors. However, no, or few, sensors were distributed within the Indian Ocean, where the epicenter of the disastrous wave of 2004 lay, as that region had exhibited a low probability of such activity. Undoubtedly, as is human nature, that region will now receive far greater sensor coverage. The fact of the matter is, the probability of a wave happening in some other, quite possibly unmonitored, area is just as high, or perhaps even greater. The dilemma of precaution here is that we can't, physically or financially, possibly monitor all possible areas of the sea. Avoidance is not a genuine option either, as those individuals who depend on the sea for their livelihood will undoubtedly return to the affected areas.

That the precautionary approach points out great difficulties with dealing with high-stakes risk is not the fault of the decision criteria itself. The dilemmas that we face arise from the nature of high-stakes risk, not from the principle of precaution. As the old adage goes, we shouldn't kill the messenger over bad news. Especially, in this case, since the messenger may have something to tell us that could greatly impact our future existence. If a water test shows the water in my well is tainted, I can't "blame" the test. The test did not cause the pollution of my water, anymore than precautionary measures "cause" the issues they bring to the fore. Now it of course makes sense to assess the logical reliability of precaution, just like we would investigate the effectiveness of any water test we might rely on. However, once we are convinced of the veracity of the "message", be it test results or a precautionary pronouncement, we have to make the best of it. That is, how do we "handle" the results. In the case of precaution, we have suggested that to make precaution workable, we may need to adjust our definition of progress. A tough pill, but one we have to swallow if we want to "cure" a case of high-stakes risk potential. Some of the backlash against precaution seems to be based on psychological factors, rather than logical ones. We don't like the message. Dilemmas, by their very nature, run against the human desire for certainty and definitiveness. Psychologically, the resistance is natural. But to affect real progress, we need to get by this distress and make some real decisions.

We have suggested that the probability of loss is difficult to assess in low probability/high-impact situations. Does the identification of increased risk presume such measurement? Well, first of all, if we apply the threshold criteria to determining possibility, we have only to determine whether the number of possible risks are increasing. Based on this threshold of possibility, it is the same as determining that as we add more tennis balls to a shoebox, the box will eventually overflow with balls. If we add more exposures to risk, as all progress suggests, we will eventually overflow their containment. Once a threshold is determined, however roughly, it becomes a matter of possibility, not probability. And as we have seen, it is much easier to determine a possibility than a probability.

If risk is in fact increasing, the fatalist finds him or herself in the uncomfortable position of idly standing by while the world becomes a more dangerous place. On the other hand, precautionary expenses (in terms or direct and opportunity costs) may serve to bring about the very catastrophes we want to avoid. Two approaches to the future of risk may help ease the tension between our choice of the two logical alternatives to how we treat high-stakes risk. The first reintroduces the basic principles of economic optimization and cost/benefit analysis to suggest that the precautionary option may in fact be the optimal cost/benefit approach as well, based on how we view "disutility" of loss. The second takes a more pragmatic view that fatalism, to some degree, is a "fact of life", and hence, rational risk management may require a *blend* of precaution and fatalism. We'll examine these two in turn.

8.2 Cost/benefit revisited

Logically, the cost/benefit framework of classical economics is unassailable. Optimization, based on "equality at the margins", is a principle that has allowed us to make great strides in the physical sciences. When the random process underlying a phenomenon are sufficiently frequent, we can still apply optimization to "expected", or probability weighted variables to achieve desired results "most of the time". Yet, we have shown that when it comes to low probability/high consequence risks, optimization based on expected values does not make sense. On the other hand, the idea remains perfectly acceptable when losses unfold in a relatively observable time horizon, i.e., when losses can be treated statistically. That the concept of optimization based on expected value should "work" at one level of risk and yet not at another seems a bit anomalous. In fact, both the precautionary approach and fatalism can be recast in an expectation framework under some very special conditions.

The precautionary approach will show the same results as an expected disutility approach if we assume the disutility of catastrophe is infinite. In terms of the disutility functions shown in Figure 2.6, in Chapter 2, the precautionary equivalent implies the function turns "straight up", to infinity, at some threshold loss value, C*. The disutility function, showing the catastrophe limit, is shown if Figure 8.2.

Any probability multiplied by infinity results in our being willing to spend an infinite amount to prevent the loss. The results will equal the minimax. We have shown the transition to infinite probability, at some catastrophe level C*, as a rather abrupt jump, or what is called a *discontinuity* in the language of mathematics. Discontinuities can present a challenge to optimization, but not an insurmountable one. We just need special rules to deal with them. Here we suggest the discontinuity be treated with a special rule, embodied in our principle of precaution. We might also have assumed the rise to infinite, or very high,

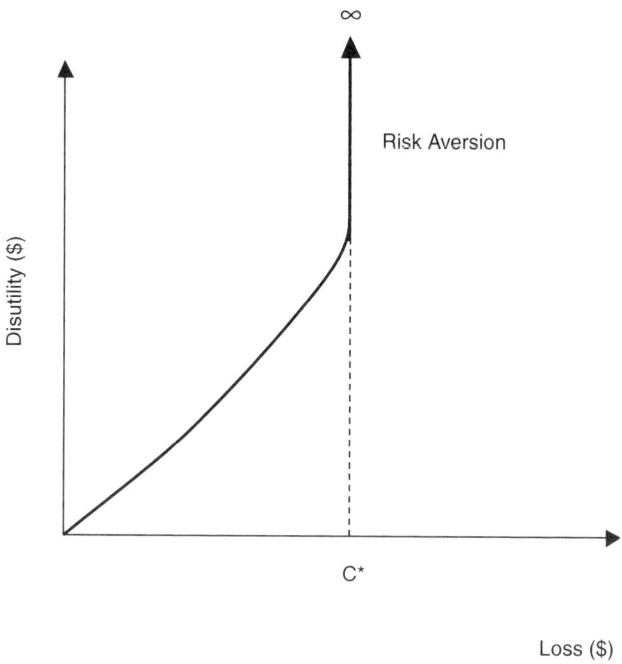

Figure 8.2 Infinite Risk Aversion (at Catastrophe Level C*)

disutility is in fact continuous, but just very rapid. In this way, we can preserve a continuous approach to optimization throughout the spectrum of losses. Under this assumption, the results of expected value cost benefit and precaution become virtually indistinguishable. Looking at it another way, given a precautionary outcome, we can always work backward to find some disutility function that would make the outcomes the same as if we used expected disutility cost/benefit. Can precautionary action be used to infer a very steep utility function? Perhaps. What this comparison suggests is that precaution can logically suffer from no more defects than an approach based on expected cost/benefit when the severity of losses is properly valued. As we have shown above, it actually has a lot *fewer* defects.

Infinite disutility implies infinite risk aversion. It makes sense that a precautionary approach would be infinitely risk averse, given the potential for what we might call infinite penalties for failure to act, i.e., catastrophic ruin, or possible extinction. The idea has similarities to the 17th century philosopher/mathematicians Blaise Pascal's argument for why one should believe in God. Known as "Pascal's Wager", it postulates that non-belief brings with it infinite penalties (eternal damnation). Even if the probability of God's existence is very small, it still makes sense to believe, based on the avoidance of infinite penalties. The argument can be viewed as an expected disutility argument (with infinite disutility), or as an application of precautionary minimax. The two are equivalent under these conditions.

The major objection to Pascal's argument is the so called "many God's" alternative. If we believe in one God, there may exist others that, though the probability of their existence is low, may also inflict infinite penalties for non-belief. Which, or how many, Gods, do you believe in? Devotion, like loss prevention expenditures, is finite. Under Pascal's argument, we face the precautionary dilemma.

As we have shown in Section 2.6, alternate behaviors with respect to risk can also be modeled by differentially weighting probabilities. An extreme application of this weighting that would give precautionary results within the expected cost/benefit framework requires that the probability of loss must always equal "0" or "1". That is, if a loss has a credible potential we treat its probability as "1". The results of cost/benefit based on expected value will now always equal the minimax result.

This transformation of probability numbers has the effect of making probability of catastrophe a very immediate thing, regardless of its true probability. In terms of uncertainty due to knowledge imperfections, it has the effect of suggesting that our interval of uncertainty always extends to "1", or very close to it. Logically, we can not exclude some very bad thing happening tomorrow. We don't think it will, or at least we live our lives like it won't, but perhaps it makes sense if, for very, very serious threats, we at least behaved like it would.

Under these modified expected value criteria (based either on infinite disutility, or a "0" or "1" weighting of probability), when the results of an expected value analysis do not agree with those of a precautionary analysis, either the consequences are undervalued, or we are overvaluing the ability of probabilities to deal with the "catastrophe problem" (averaging is ineffective when losses get large). In fact, many debates over the treatment of high-stakes risk center on exactly what, or how much, is at risk. The greater we value the loss, the more conservative our expected value decision, probabilities being "fixed". As should be obvious by now, sufficiently large loss potentials (or, disutility) can overwhelm any probability, no matter how small. How do we properly value human life, or can we? Is its value, in fact, infinite? How about a rain forest, or an animal species? Proper valuations would themselves make cost/benefit decision based on expectations much more conservative (i.e., closer to precaution).

The transformations we have suggested to expected value/utility decision-making might seem trivial. Indeed, there is no need for a complex apparatus of probability estimation and calculation under these transformations. In the limit, probabilities once again become irrelevant. A simple application of minimax will suffice. Direct application of the minimax also avoids the idea of averaging of results, which we have shown does not make sense when probabilities are small and loss potential very large. This all suggests that precaution is not just some variant of expected value decision-making. Yet these transformations do shed additional light on the ideological underpinnings of precaution. With respect to high-stakes events, we behave "as if" the probability of loss is certain. Our valuation of extinction, on an individual, business or societal level, implies infinite disutility, and hence risk aversion.

Let's compare similar modifications with respect to fatalism. We could always achieve fatalistic results in an expected value analysis if we assume the probability of catastrophe is "0". This explains why some refer to what we have called the fatalistic approach as "optimistic". Optimism assumes the catastrophe will never happen, so we need not do anything to try to avoid it. Alternatively, we could modify disutility to equate all losses, perhaps at "0". This reflects a "nothing to lose" attitude that might also be consistent with "doing nothing".

Fatalism forces a sort of delusional approach to catastrophic risk: It doesn't exist. Or for all intents and purposes, we choose to treat it like it doesn't exist. Either that, or we don't care, or perhaps we are forced by circumstances not to care, about serious consequences. The paradox in the first case is that catastrophes do happen. We know they exist... yet we must treat them like they don't. We make significant expenditures and take great care to reduce statistical risks, such as an automobile "fender bender", or the prevention of a winter cold, yet take no precautions against potential catastrophe.

Clearly, we would all live a precautionary life if precaution was "free". There is plenty of evidence to suggest that precaution is the preferred approach to danger when costs are modest. Enter, the proportionality principle. We are willing to spend some reasonable amount on precaution. "Reasonableness" implies a rough cost/benefit comparison, though expected values and other probabilistic criteria do not enter. When costs of precaution get high enough, we face a dilemma.

The costs, including those of foregone opportunities, may themselves ruin us. The resulting discomfort forces us to seek alternatives.

Attention to alternatives early on in the process of defining progress may in fact be based on the recognition of the true, possibly infinite, cost of catastrophe in the wider cost/benefit framework. This effort becomes the focus of precautionary oriented science. By seeking alternatives we try to prevent precautionary dilemmas. In this sense, alternatives analysis is ultimately an "economical" response to high-stakes risk. Its focus is distinctly positive: What can we do to mitigate high-stakes risk *before* they become a problem?

On the other hand, the shortsighted approach to post-fact economic optimization ("risk management") focuses on what we *can't* do. Since we don't have infinite resources to expend on safety, we need to ration resources accordingly. We can make automobiles completely safe, reduce fatalities to zero, if we spent the entire United States GDP on auto safety. But such an expenditure would obviously result in a more serious catastrophe. And who would pay for a car that costs twice or three times as much, because of safety requirements?

Viewing economics from the perspective of scarcity places the emphasis squarely on production, or "supply". The proper focus is on human beings. Economics is the satisfaction of human needs, wants and desires. Among them, safety. Resources are not unlimited, to be sure. We have to realize, however, that resource limitations are a factor of how ingenious we can be. Our ultimate goal should be to expand the use of these limited resources to the fullest. Utopia exists when these restrictions no longer limit our ability to satisfy our needs. We should strive for this ultimate goal, not attempt to justify "second choices". Cost/benefit is too often cast in a static vein. It shouldn't be "do the best with what you have", it should be "do the best". Challenge the frontiers of the possible. We don't need to settle for an existence that forebodes catastrophe, we need to work against it.

In the perfect world, we don't need to build cities in flood prone or earthquake prone areas. We don't need to accept the benefits of life saving drugs that carry with them the potential for catastrophic allergic reactions. The modified food sources we rely on in the perfect future cannot have the potential for making us ill or killing us. Why do we strive for this perfection, yet assume that all progress toward it brings with it the potential for serious risk? There is no natural law that says to make progress we must accept the risk of catastrophe.

Ultimately, cost/benefit, and the economic optimization it implies, requires we live the most beneficial life we can, while expending the resources to do so wisely. This premise need not presume some exact calculus by which to "figure" the results. The desire for such an unattainable calculus is what leads us to accept mathematical constructions for dealing with all phases of life, that though they make perfect mathematical sense, do not conform with reality. We don't need such a calculus to tell us who or how to love, respect, trust, or value. And we don't need such a calculus to tell us what we should or should not be afraid of.

8.3 Reconciling fatalism and precaution

As strong as the logical arguments for precaution are, it would seem that at least some degree of fatalism toward high-stakes risk is inescapable. We have already established the fact that for any sort of precautionary approach to be workable, we must at least establish some non-zero probability threshold for what we call "practical impossibility". Acceptance of exposures below this threshold implies at least some degree of fatalism. So even the most ardent precautionist requires some fatalistic tendencies. We have proposed a reasonable guide for this threshold as being the natural "background" level of risk found in nature. This background level of risk becomes, in effect, the goal of precaution. Once again, nature is our guide. Nonetheless, if a "man-made" peril is associated with practical impossibility of significant adverse consequences, this peril itself blends against the natural background. It becomes part of an uncontroversial fabric of nature and man, existing "side by side" as it were, with the resulting benefits.

It is also obvious that people display a degree of fatalism towards catastrophic risk that may be quite possible. We "accept" risks that quite obviously fail the precautionary test. Given the number of fatal auto accidents per year in the United States, we would be hard pressed to say the individual risk of driving is negligible. Yet we all (most of us) drive. Five million people live in the greater Los Angeles area, despite the credible threat of a major earthquake in that region. Precautionary evaluation might very well suggest otherwise.

As we suggest throughout this work, the fatalistic approach remains the dominant approach to high-stakes risk in the world today. Precaution is certainly practiced, but often only to the extent it is relatively inexpensive, in proportion to the activities at hand (the "proportionality principle"). There are also decisions made on the basis of statistical expected value criteria in what is clearly the non-statistical domain of catastrophic loss. Some of this application is undoubtedly out of ignorance. In other cases, it is simply based on a desire to promote self-interest with misleading, yet "scientific looking", results. Whatever the case, expected value decision-making (and its variants) should properly be considered a case of disguised (intentionally or unintentionally) fatalism. That said, the publicity surrounding the application of expected value cost/benefit methods suggests these methods are used far more than they really are. In organizational risk management, health and safety, environmental initiatives, and even corporate planning, formal decision-theory has found little practical application. In a 1970 issue of the Harvard Business Review, decision theorist Rex V. Brown asked, "Do Managers Find Decision Theory Useful?". The answer was, "no". Many decisions theorists attributed Brown's results as growing pains, but 30 years later the answer is still the same. In the field of risk assessment most formal risk studies are still most used in public forums, where there is at least an element of controversy surrounding the decisions under study. The use of decision theoretic methods, based on classical economic optimization and expected value criteria, for more individualized, "internal" decisions by individuals, businesses and communities remains fairly minimal. That leaves, once again, the tough choices: Precaution or fatalism.

If fatalism exists, and if fatalism is fairly widespread, what might we infer from its existence? If fatalism exists side by side with precaution, what characteristics of the exposure to catastrophe might differentiate the two? In fact, we might observe that some of the risks "accepted" under fatalism are more troubling than others. Acceptance is not so much reasoned, as forced. We might call this variety of fatalism "acquiescence" to risk. Operating under conditions of a forced exposure to things we fear, but have no choice in, presents a particularly dismal view of human response to risk. Fatalism is not resignation, it is reasoned acceptance. The former is a passive response to risk, the latter is active.

As we said early on in this book, risk management is ultimately about reducing our concern and worry over risk. In the statistical treatment of risk, we do so by making sure things average out in our favor. In the realm of catastrophic risk, we do so either by applying precaution or adopting a reasoned fatalism with respect to risk. Mere acquiescence, however, does not remove worry.

The wider goal of risk management may be to *eliminate mere acquiescence*. Of course, one way to do so is by working toward the implementation of suitable precautionary alternatives. Yet, might not some features of risk provoke a "reasoned fatalism" (genuine acceptance, versus acquiescence)?

In fact, studies of individual decision-making behavior have shown that individuals routinely display a greater tolerance, or "acceptability", for risk, even potentially catastrophic ones, to the extent that the activities that promote such risk posses certain qualities. These factors in risk perception might give us some indication why some precautionary dilemmas concern us more than others.

Among those factors frequently noticed in studies are the degree of fairness or equity, perceived degree of control over the risk, and the degree to which the exposure to risk is voluntary. A more complete list is shown in Figure 8.3. These factors influence risk acceptability, and would therefore suggest that there might be some differential application in precaution versus fatalism. Those who engage in the recreational sport of mountain climbing certainly expose themselves, knowingly and intentionally, to risks that present the distinct possibility of catastrophe. While minimax would suggest precautionary avoidance, a distinctly, and sometimes quite overtly fatalistic stance is taken. If it happens, it happens.

If we examine the list of factors that affect how risk is perceived, we notice that many of the factors pertain to what we have called "natural" risks. It might be further argued that in an unfettered state of nature, we would have none of these concerns. That is, risk acceptance would be voluntary, fair, understandable, and so on. A reasoned hybrid approach to catastrophic risk may exist, where we are fatalistic toward "natural" risks and precautionary toward man-made ones.

In this framework, comparison of risk in terms of probabilities and impacts becomes irrelevant. An environmentalist may enjoy a kayak ride down the rapids of a river, entailing the distinct possibility of serious injury or even death. This suggests a fatalistic approach to the

- Fairness/equity
- Familiarity
- Degree of control
- Voluntary vs. involuntary
- Degree of trust

Figure 8.3 Factors Affecting Risk Acceptance

danger. At the same time, he or she might oppose the construction of a nuclear power plant along the banks of that same river, despite the facts that the threat from the power plant may be strictly lower in terms of probabilities of harm or fatality. With recreational kayaking, risk acceptance is voluntary, the exposures understood, any adversity is distributed fairly – in a word, risk acceptance is "natural". On the other hand, nuclear power presents an exposure that is not well understood, the possibility exists that some may gain (power companies) at the expense of others (the local public), and reasonable precautionary alternatives may exist.

Lennart Sjoberg, head of the Center for Risk Research at the Stockholm School of Economics has suggested that we are moving towards a conception of risk perception as ideology rather than emotion. This suggests that risk perception has an objective basis, founded, perhaps, in the "naturalness" of risk. Risk acceptability follows in that the more natural the risk, the more acceptable. This all suggests an ideological approach that determines the blend of fatalism and precaution in our response to catastrophic risks.

If we further examine the reality of fatalism, we see that while a hybrid approach based on naturalness makes sense, people do react fatalistically to some man-made perils as well. It might be that certain man-made exposures to risk could in some fashion "blend" with the natural environment. In a way, they have become part of a natural course of progress in humankind. Driving an auto may be a case in point. The individual risk of driving presents the distinct possibility of serious accident or even fatality. At first glance, driving may be a candidate for precautionary avoidance. In the modern world, such avoidance can have serious negative consequences. Not driving could impair ones livelihood, and quality of life, adversely. A precautionary dilemma presents itself. Again, we are doomed if we do, and doomed if we don't. Most people do drive, and as such, this action represents a degree of fatalistic acceptance of the (possibly catastrophic) consequences. Why? We might suggest that a reasoned acceptance of the risk has been granted based on the perception of driving risk. The risk is fairly "known" and understood, and there is a certain equity about those exposed to the risk. It is for the most part a voluntary effort, though it may be argued that pursuing a livelihood in the world today demands it. There seem to be no particular trust issues, although again, we might question the complete assessment of alternatives based on the existence of an automotive industry that is one of the largest in the world. Overall, however, we might classify driving as a somewhat natural part of human life. That this reasoning is based on acceptance rather than acquiescence can be gauged by our reactions to driving "catastrophe". We are always shocked and remorseful when we hear of a driving fatality or serious injury. Yet, we don't often hear reactions of the type, "They should have never even learned to drive a car!". The absence of the activity, despite its risk, does not seem to be an option.

Consider on the other hand, the exposure to a chemical pesticide to fight infestations of possibly diseased mosquitoes. In this case, the properties of risk perception are not favorable. The toxicology of chemical agents is not well understood. Carcinogenic properties can be detected, indicating a potential for risk. Whether this potential qualifies as presenting a genuine possibility of risk, say, against some natural background level, is difficult to determine. The exposure is to some extent voluntary, except in the case where pesticides may be applied by the community. Overall trust may be an issue, as pesticides are sold by companies that are in it to make a profit. What are my reasonable alternatives? Are there biological solutions to the issue, such as the introduction of natural mosquito predators? Have all the prevention options been considered, including eradication of mosquito breeding conditions? If fatalism prevails, it will probably be based on acquiescence. The dilemma may be more reasonably approached via avoidance.

"We need to take risks to succeed", "you can't be afraid of everything". These statements all suggest a fatalistic attitude toward risk. In some cases, this attitude is applied to man-made perils with catastrophic potentials, with relatively little trepidation. Candidates may include driving and automobile or working with electricity. On the other hand, we may not be so quick to take a purely fatalistic stance to things like nuclear power or genetically modified foods. The distinction may be that some man-made perils "take on" the traits of their natural counterparts. They may become "essential to life", while their risks are, or are becoming, known in an honest fashion. Any potential adversity is equitably distributed, and decisions participatory (i.e., democratic). The "winners" and "losers" in the risk debate become harder to distinguish.

8.4 Shaping the future

In assessing any high-stakes risk potential, we begin the process by ascertaining the "possibility" of risk based on what we know of the likelihood for bad things happening (however vague). If risk is possible, and catastrophic, we exercise precaution. This, in its most simple form, is the principle of precaution we must practice if we are to successfully deal with catastrophic loss potentials.

We can assess all exposures to risk, natural or man-made, with an eye toward precaution. The best risk, is no risk – all things considered. Often, precautionary risk management is "low cost" or "no cost". Other times, it may require us to adjust our ideas of progress, and think in terms of alternatives. Our course of action is based on our desire to keep in harmony with a nature against which has played a rather long streak of evolutionary survival. The goal is always to limit possibility of disaster below some "de minimis" threshold, based on natural background risk.

Sometimes, we face dilemmas in that precautionary action based on the principle of minimax'ing losses results in costs that may in fact approach the cost of the catastrophe we are trying to prevent. In these cases, we face a dilemma. When diligent search for reasonable alternatives proves futile, where do we go from there? If the exposure is sufficiently "natural", that is, if the criteria for risk acceptance are fairly met, then we really don't have a dilemma after all. We become fatalists by choice, rather than being forced into the position out of despair. Cognitively, this whole process would seem more satisfying than using fanciful applications of cost/benefit analysis.

To the extent we can operationalize the factors affecting risk perception, we can move risk acceptance from the subjective to the objective. We don't accept risk because we "like it". Nor does this imply that risk acceptance should be determined by a simple "vote" by all those potentially affected. Risk assessment should be a democratic process, in the sense that all involved should be informed and able to give input, as desired. This does not mean that ultimately the majority rules, depending on subjective likes or dislikes. This is as dangerous as making decisions based strictly on immediate financial interests in a risky technology or activity. You can't fool all the people all the time, but they might just be fooled (or misinformed) long enough to cause some irreversible damage.

A considerable amount of risk management proceeds, and will continue to proceed, on the basis of statistical cost/benefit analysis. When random adverse events manifest themselves within the relatively short-run, say maybe 10 to 25 or so years, the results "average out" within a fairly perceptible time frame. Statistical arguments in this realm remain verifiable, and hence provide a credible basis for decision. Such statistical techniques, however, are simply not applicable to high-stakes risk.

Methods that guarantee successful decisions in the "long-run" don't make much sense if in the long-run we cease to exist. In terms of the likelihoods of "typical" losses in a category, we may therefore partition the probability dimension as shown in Figure 8.4.

To make decisions when the stakes are high, we need to consider the somewhat diametrically opposed alternatives of fatalism (risk acceptance) and precaution (avoidance). The decision we make when faced with high-stakes exposures is between precaution and fatalism, not precaution and expected valued based cost/benefit. We summarize the "pros and cons" of the high-stakes decision options in Figure 8.5.

While we may be driven to some sort of statistical analysis of catastrophe just to show we are doing something, statistical methods in this regard are *not* "better than nothing". Not only is expected value decision-making useless in the case of catastrophic exposures, it can intentionally or unintentionally lead us to distinctly bad decisions. Accepting catastrophe based on fruitless expected value calculations, especially when reasonable alternatives may exist, is a bad decision. By recognizing the true alternatives, we eliminate the power of scientific sounding methods, like expected value optimization, to mislead.

The lot of those that promote risk assessment based on precise estimates of probability and expected value decision criteria would be much better off if they could at least sometimes say "We prevented the terrorist attack!", "We prevented great loss of life and property damage

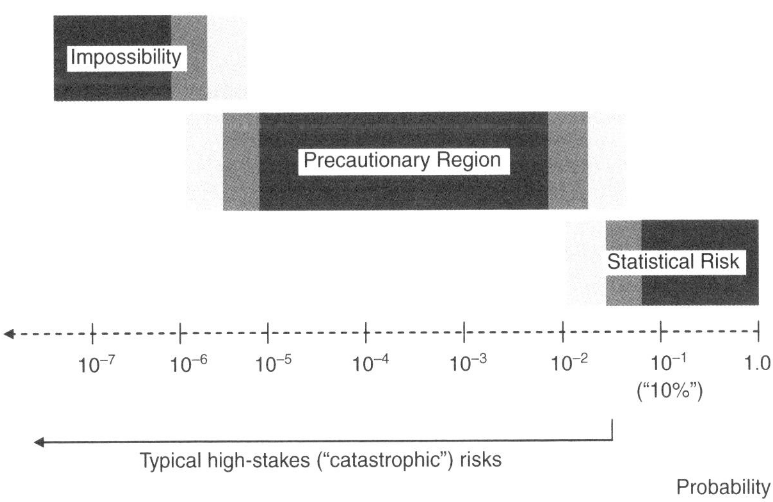

Figure 8.4 Partitioning the Probability Dimension

Approach:	"Pros"	"Cons"
"Fatalism"	1. Easy 2. Low cost/no cost 3. "Philosophical" justification	1. Ignores disaster potential (?) 2. Can be manipulated by "special interests"
Precautionary Avoidance ("minimax")	1. "Eliminates" risk	1. Expensive 2. Hinders (or at least requires redefinition of) "progress"
Expected Value	[NONE]	1. Promotes "disguised" fatalism

Figure 8.5 The "Pros and Cons" of High-Stakes Decision Criteria

caused by the hurricane!", or "We prevented great loss of life due to the chemical release!". The fact is they remain powerless in the face of such natural and man-made calamities. After the disastrous event has occurred they redouble their efforts to convince us how important expected value criteria are, exposing what can only be described as much too little, much too late. Unfortunately the shock of such disasters often leads people to seek comfort by those that promise some degree of relief from future catastrophes, no matter how tenuous the reasoning behind such approaches.

Shaping the future of risk may involve identifying some blend of risk avoidance and acceptance (precaution and fatalism), or if any proper blend exists at all. The responsibility of professional risk managers will be to help identify this balance. Risk management is ultimately a practical endeavor, not a theoretical one. This at the very least requires an open mind towards all possible risk management techniques, and a full understanding of all relevant approaches available. This is a little bit of what we have tried to do here. So, if anything, this book is just a starting point or a jumping off place, suggesting guidance in the process of making the world a little better place.

If it is ultimately the precautionary approach to risk we choose, we need to make our commitment to it explicit. A concise statement of the precautionary principle, articulating the points we have discussed in this presentation is contained in Appendix C, below. Such statements can be incorporated into the entity's risk management *mission statement*, or charter. Of course, precaution will not encompass our entire risk management mission, although it is arguably the most important component. Statistical risks may be addressed by including statement to the effect that, where applicable, we will use methods to "minimize the cost of losses" or "make cost effective decisions with respect to risk". The problem is that most mission statements with respect to risk rarely go beyond those that address statistical risk. The implication is that catastrophic losses are to be treated statistically (via expected value cost/benefit), or that we choose to remain silent on our approach to catastrophic risk, which more often than not turns out to be fatalistic.

8.5 Can we get there from here?

Given the entrenchment we face in modern technologies, and even what we might call our way of life, does progress toward a more precautionary stance seem possible? In other words, *can we get there from here*? At the very least, we need to make room for some degree of fatalism (reasoned risk acceptance) based on the naturalness of exposures. Beyond that, the enormous sunk costs of human history make precautionary dilemmas enormous. Consider the human population living in Southern California. By all indications, a catastrophic earthquake with enormous loss causing potential, both in terms of human life and property values, is certainly within the realm of the possible. No precautions in terms of the engineering of current structures could significantly reduce the impact of such an event. The only sure fire prevention is to move away. Hardly what we would call a viable option. What comfort to be had in such a situation is that the event would be, to a great extent, natural, with the acceptance of risk, for the most part, voluntary.

This is not the case with many man-made potentialities for disaster. In innovation, we are not bound by the sunk costs of history, and perhaps have more room for genuinely precautionary action. The informed choice of action depends a lot on science (and how we choose to use it). In many cases, the history of man-made inventions is short enough that we may be able to alter its future course at a relatively small cost. Automobiles with hydrocarbon powered internal combustion engines are a case in point. Exercising that discretion is up to the community, and up to businesses that provide for that community to step up to their social responsibilities.

Perhaps where we are at now is not an accident of history, or a failure to properly manage high-stakes risk. Maybe it is where we were meant to be. If the dominant mode of thought with respect to high-stakes risks is fatalistic, we see around us the product of a fatalistic world... and we live with it. If this is not what we bargained for, then it may or may not be too late to do something about it. Sometimes to start fresh you have to see the previous regime through to its conclusion, whatever that may be. In the middle of it all, we may start to look for some easy way out. Extrapolation of statistical techniques is one such panacea: Let's apply some prevention to the threat of catastrophe, but only where it makes "economic sense".

Unfortunately, progress towards a fully precautionary solution to the catastrophe problem cannot be made piecemeal. That applies on the individual, business and community levels. Logically, as long as one source of preventable catastrophe exists, the catastrophe problem continues to exist in full force, threatening the existence of the individual, entity, or society. Prioritization and other half way measures may ultimately be of little value. Relativities among catastrophes are irrelevant: When you're gone, you're gone. It doesn't matter if you are run over by a 300 pound rock, or a 3,000 pound rock. Proponents of precaution who suggest such piecemeal solutions, or gain comfort from their adoption, are deceiving themselves to a degree no less than those that espouse partial measures based on expected value cost/benefit.

The logical incongruity of piecemeal solutions to the catastrophe problem does not mean that recognition of the issue might not proceed in a gradual fashion. We may come to realize with time and reflection that "we have to do something, and we have to do it now". How far off, or even *if*, such realizations are forthcoming we cannot say here. A lot depends on our perceptions of catastrophic risk. As we noted above, fatalism might be a legitimate alternative to the extent that we can say we have achieved reasoned acceptance, as opposed to mere acquiescence. At what level these perceptions surface at the individual, business or community level is uncertain.

The existence of precautionary dilemmas should never, in and of itself, become an excuse for fatalism. Neither should these dilemmas drive us to use statistical methods in an area where they are inapplicable. If anything, the existence of precautionary dilemmas should drive our ambition to do something about them. We would suggest that the first step is a more reasoned analysis of decision-making options under high-stakes risk, unclouded by special interests (if that's even possible). We have summarized some of the arguments here, and recommend other pertinent references (see Appendix B). The rest is up to the reader.

Appendix A A Working Model of Precaution

In the discussions above, we suggest that the *precautionary principle* is a logical criterion for decision when the stakes are high. While articulations vary, we see that most boil down to the simple rule that possibly catastrophic events must be avoided. We deal with high-risk exposures by assessing their properties, and applying this precautionary principle, or rule.

Precaution seems intuitive. Nonetheless, critical debates arise concerning the alleged "vagueness" of the precautionary principle as a criterion for dealing with high-stakes risk. Often, multiple linguistic variations are cited as evidence that the principle is ambiguous, and hence, unworkable. Most differences, on closer examination, prove merely semantic. In fact, most doubts about our ability to make the precautionary principle operational arise from a lack of understanding of uncertainty that results from imperfect knowledge.

As we have shown above, when uncertainty due to knowledge imperfection exists, the best we can do is to identify our world imprecisely, using approximations. We will introduce here a way to more accurately represent these types of uncertainties, using the theory of *fuzzy sets*. Fuzzy sets can be used to operationalize the linguistic rule we call the precautionary principle. Once we have achieved this formalization, we can proceed to build a working model of precaution that lets us apply the idea to real-world risk management problems.

A.1 An introduction to fuzzy sets

As we have described in Chapter 2, in dealing with real-world risks, we face an uncertainty due to knowledge imperfection, in addition to the uncertainty introduced by randomness. Knowledge imperfection manifests itself in our inability to distinguish among alternatives with precision. When our knowledge about some property or event are imperfect, we can only specify model parameters and measurements as a range of possibilities. As we have shown above, the simplest way to describe this uncertainty is using some "interval of possibility" along the scale of potential outcomes. However, we can often associate varying degrees of confidence, or credibility, to outcomes within the range. This "graded nonspecificity" is known as fuzziness. We can apply fuzziness to the basic building blocks of logic, the set.

Human knowledge is organized around concepts – like *risk*. Concepts can be conveniently formalized using sets. A set is simply a collection of items (called elements) that share some property or properties. Concepts are rules, or definitions, used to classify objects or observations into sets based on these properties. We use the resulting sets in a variety of reasoning tasks. For example, we may include them in rules, developed through experience: If observation X falls in set Y, then take action Z.

In traditional logic, set membership can be precisely defined: Either something belongs to the set, or it does not. We call such sets *crisp*. This view of sets, which implies a high degree of knowledge about our subject, underlies much of modern science. In geometry, the concept named triangle is defined as a closed figure with three sides. If a figure has any more or less than three sides, it is not a member of the set of triangles.

When fuzziness due to knowledge imperfection enters, a precise definition of sets is not feasible, or desirable. Elements can only be defined to various degrees. Graded set membership is the basis of the idea of fuzzy sets. A fuzzy set is a set in which inclusion need not be defined as "all or nothing". Fuzzy sets admit a degree of belonging according to a concept known as *membership*. Membership expresses this degree of belonging on a scale of zero to one. The boundaries of a fuzzy set correspond to a simple interval of uncertainty. A membership of zero means that an object or observation is definitely not an element of the set. A membership of one means it is a "fully possible" member. Numbers in between show degrees of belonging. By admitting degree, fuzzy set membership lets us more accurately express our uncertainty than a simple interval of possibilities (all with a membership of one).

Consider the concept of tall men and it's set. Figure A.1.a shows a fuzzy definition of the set, using a membership function. The membership function is an equation that relates fuzzy membership degree to potential members of the set. The potential members are known as the universe, or universe of discourse.

For example, the concept "tall" as it applies to men, shows that men less than 60 (five feet) inches are definitely "not tall", while those greater than 72 inches (six feet) definitely are. Heights in between fit the concept tall to various degrees. Notice that exact cutoffs do not properly represent vague concepts like tall. For example, we could set some arbitrary limit of 72 inches as tall, anything under that is not tall. So a 71.9 inch man is not tall? Exact cutoffs require precision in both measurement and use of a concept. Precise measurements are not feasible in many real-world situations. Above all, such exact cutoffs require that we know far more about our world than we really do. The relationship between height and basketball prowess, for example, is simply not reducible to precise computation. When selecting our team, however, we want to make sure we have plenty of tall players.

The concept tall is a threshold concept, shown by a sort of "s" shaped membership function that we call a shouldered function. Fuzzy sets can also be unimodal, as shown in Figure A.1.b for the concept "cool" with respect to springtime temperatures in New England. Temperature lower than 35 degrees and higher than 65 are not what we may call cool (they might fall into the regions "cold" and "warm", respectively). We could therefore express the concept cool using an interval in between. However, intervals have exact boundaries too. A more natural representation of the approximation cool is to use gradual borders.

Note also that the spread of the membership function provides a natural measure of the degree of uncertainty involved. At the extremes, a single point with a membership of one (all other observations have a membership of zero) implies perfect knowledge, such as that gained through mathematical definition: The number of side in a triangle is three. On the other hand, complete ignorance is defined a membership function in which all observations (potential elements

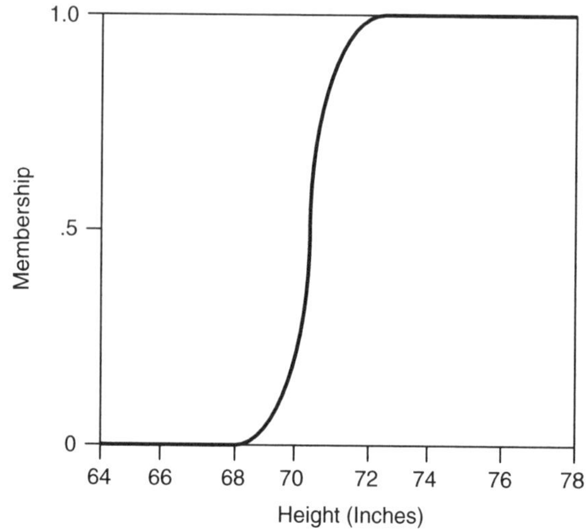

(a) "Tall" (Shouldered)

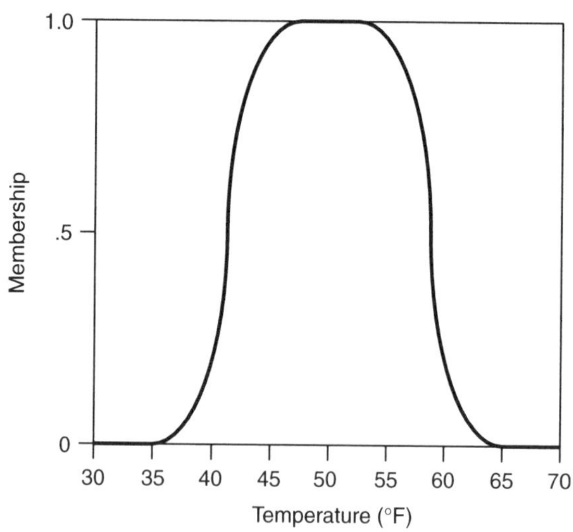

Figure A.1 Fuzzy Membership Functions

of the set) have a membership of one – when we don't know, everything is possible. Our knowledge of most concepts usually fall somewhere in between.

Fuzzy set membership functions are usually obtained by direct questioning of human subjects familiar with the concept. Though obtained from the experience of the individual, memberships are not subjective in the sense of being peculiar to that individual. They have a link to objective reality. That link is instrumental, in the sense we judge memberships by how well they let us reason about a complex and uncertain world. If we use concepts with too much precision, we will find they cause inaccuracies and inconsistencies in reasoning. We then, instrumentally, relax our definition.

In specifying membership functions, there is always a tradeoff between specificity (and hence informativeness) and truth. By making specific statements under conditions of uncertainty, we risk that they may be false ("the temperature outdoors tomorrow will be 82.5 degrees"). On the other hand, non-specific statements may become vacuous, providing no basis for action at all ("the temperature tomorrow will be between 0 and 150 degrees"). Fuzzy sets are constructed so as to strike a balance between specificity and truth, based on available knowledge.

A.2 Formalizing linguistic rules using fuzzy sets

Though imperfect, approximations based on fuzzy sets can provide useful information for decision. Consider this simple rule-of-thumb: "If it is *cold* out today, then don't wear shorts". Vague, yet useful. The vagueness is not in the rule itself, but rather in the properties we attempt to measure and their theoretical relationships. The complexity of these measurements and relationships does not permit us to define the rule any more precisely, without ruining its applicability. The same idea applies to linguistic principles of precaution: "If an activity entails the *possibility* of significant adverse consequences, then avoid the activity". In applying the precautionary principle, we first determine that the event in question is in fact "possible". Then we apply the *minimax* (loss) criteria for decision under uncertainty, which suggests that we choose actions so as to minimize the maximum possible loss.

To trigger the minimax, precaution requires that an impact be both possible and *significant*. For simplicity of exposition, we will treat only the possibility dimension of high-stakes losses here, assuming that the consequences we are dealing with are unequivocally catastrophic. The analysis is easily extended to the consequence dimension.

Clearly, setting a threshold of "zero" probability for impossibility is too strict for practical application. The statistical laws of physics tell us that the there is some non-zero, albeit tiny, probability that air molecules can act so as to re-inflate a flat automobile tire. No one would sensibly wait for them to do so, in lieu of calling for a tow truck. Instead, we can rely on some very small probability threshold value to represent "practical impossibility". Given the complexity of applications of this principle, and their related uncertainties, it does not seem reasonable to suggest any precise threshold. Rather, we can represent impossibility by a range of probabilities. Figure A.2 shows an articulation of the fuzzy concept of "possibility" in term of a fuzzy set.

For the concept of "possibility" as we have framed it, the universe of discourse is probability numbers on the scale 0 to 1. To more conveniently reflect low probability events, we will use a logarithmic scale, with the realm of interest being annual probabilities from "one in a thousand" (.001, or 10^{-3}) to "one in a billion" (10^{-9}). The scale in the figure is marked by factors of 1/10, i.e., 10^{-x}.

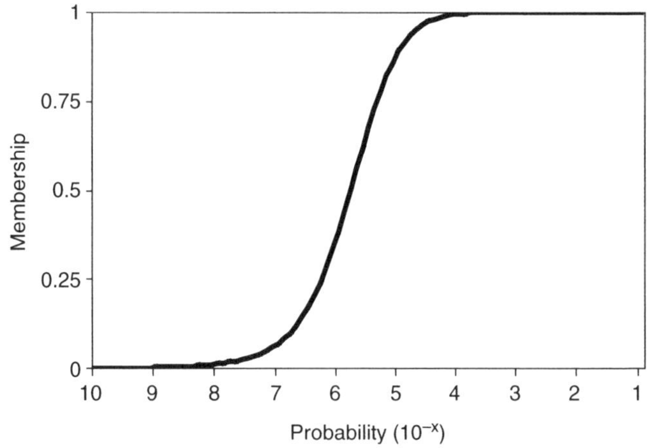

Figure A.2 A Fuzzy Definition of "Possibility" (i.e., "Possible Risk")

Probabilities in the range 10^{-3} to 10^{-4} are clearly "possible". Likewise, probabilities below 10^{-6} (the proverbial "one in a million) or so have been traditionally considered outside or nearly outside the realm of possibility. The dividing line between possible and impossible is, however, fuzzy. We represent this by giving the probabilities "in between" grades of membership between 1 (fully possible members of the set of probabilistic "possibilities") and 0 (full non-members of the set "possible, meaning "impossible"). Measurements in between show various degrees of membership, representing the uncertainty we feel in specifying any of these borderline numbers as full members of one set or the other. Another way to look at these uncertainties is to say that probability numbers in this zone have mixed properties of both "possibility" and "impossibility" to the given degree. This idea clearly has different implications that saying a probability is definitely "possible" or definitely "impossible". By removing this basic principle of "two valued" (belongs/does not belong) sets, we have to introduce a modified logic for dealing with fuzzy sets, known appropriately as *fuzzy logic*. Fuzzy sets and their related logic have a well-developed (and still developing) scientific background. They have been used not only in social sciences and human decision-making, but also in many practical applications in the engineering and control of physical systems.

A.3 Fuzzy measurement

We must recognize that our assessments of probability of real-world events will themselves be fuzzy, adding another dimension of uncertainty to the application of precaution. We may have more knowledge about some probabilities than others, making their membership functions narrower around some "best guess" estimate. The wider the membership, the more uncertain we are about the probability. In fact, we can use measures of the spread of these fuzzy memberships measures to express the degree of uncertainty involved in our estimates.

Our fuzzy set interpretation of the precautionary principle can now be fully operationalized. To match the fuzzy definition of possibility to related measurements, we use the concept of fuzzy intersection, or degree of overlap, between fuzzy probability measurement and "possibility". The degree to which the two memberships overlap determines how well a fuzzy probability measurement "fits" the idea of "possibility".

Several different hypothetical (but realistic) "measurements" of probability are combined with our definition of possibility (Figure A.2) in Figure A.3. In (a) we show the case of relatively low uncertainty about a probability associated with some event. While uncertain, the membership falls, for the most part, outside of our fuzzy threshold for possibility. Events with profiles similar to (a) are not candidates for precaution. One example would be the process of drinking water. Not overtly harmful, and indeed quite necessary to life. Or consider the occurrence of a major earthquake in southern Florida. Southern Florida is an area where both existing records, and scientific analysis of its geology, suggest an absolutely tiny probability of a major (severely damaging) earthquake. On the other hand (b) shows some modest uncertainty as to probability, but most of this uncertainty falls within the envelope of our definition of possibility.

The event in (b) is fuzzy, but "possible". Precaution requires we avoid event (b). Examples include the unprotected use of carcinogenic substances whose properties are fairly well understood, such as asbestos or benzene.

Example (c) shows the case where fuzziness complicates the analysis. There is significant uncertainty as to the probability of event (c), as evident from the "spread" of the membership function. Clearly, our knowledge is not sufficient to exclude the "possibility of possibility", and hence we would treat such events with precaution. Such uncertainty is typical of complex technological risks with the potential for highly adverse impacts. One example is the emerging science of genetic modification of food crops to increase yield and reduce damage from pests. The long-term effects of crops like these on humans and their environment is not well known. There are suggestions, however, that harm in the form of allergic or other poisonous reactions may be attributable to such crops.

Allowing for uncertainties in loss potential is a unique feature of the precautionary principle. It is this feature of precaution that suggests that highly uncertain events (i.e., those that can not be excluded from possibility) be avoided. It is often this feature that is the most controversial as well. Again, the problem is not with the definition of precaution itself, but rather with the refusal of many to admit such uncertainty exists. Where controversy exists, the specification of fuzzy membership functions lets us bring this controversy to the fore, thereby focusing the debate. Irresolvable controversy among experts is in fact a sign of genuine uncertainty.

Critics of the precautionary principle often attempt to force the fuzzy definition into a precise model. The threshold probability is set at a fixed probability, often assumed to be strictly zero. By equivocating over the threshold (when in fact no exact threshold exists), or by assuming that the only defensible threshold is "zero", these critics often attribute to it the finding that "everything is risky", and hence "everything should be avoided", leading to a paralysis of action.

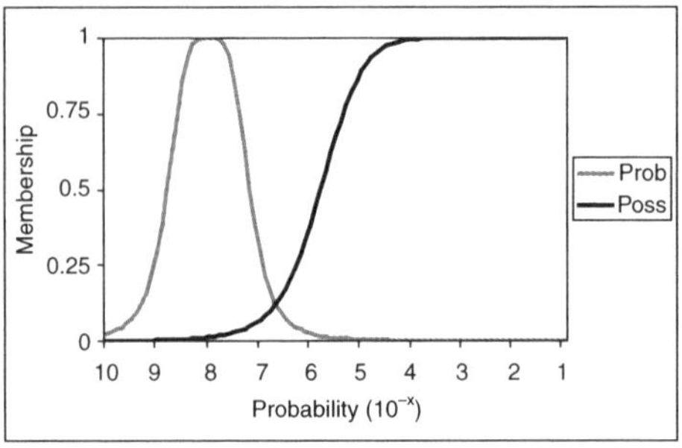

a) No precautions, low uncertainty (e.g., major earthquake in Florida)

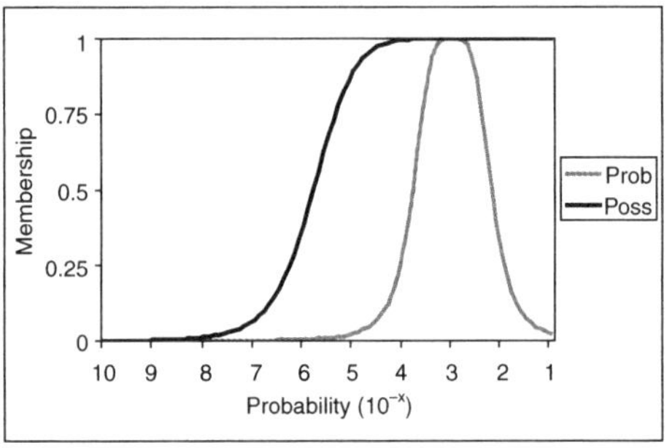

b) Use precaution (avoid), low uncertainty (e.g., asbestos)

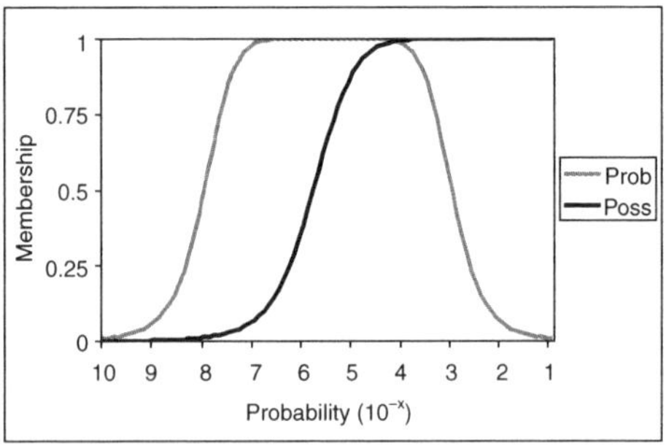

c) Use precaution, high uncertainty (e.g., genetically modified crops)

Figure A.3 Applying the Precautionary Principle

Note that intuitive application of the linguistic definition in our day-to-day lives results in no such difficulties.

Likewise, by assuming that measurements of probability are always made with precision, we ignore that critical component of the principle that accounts for uncertainty due to knowledge imperfection. In the linguistic articulation of the precautionary principle, possibility is taken as a call to action, recognizing that the need for scientific certainty, or certainty about probabilities, should not preclude such action. Since the precise framework of probability cannot account for epistemic uncertainty within risk analysis, action based on such uncertainty is completely excluded. Simply the wrong tools are being applied, as if we were using a hammer to cut a sheet of wood, or a saw to drive a nail. Not surprisingly, frustration results.

A.4 Building a working model

A fuzzy interpretation of precaution lets us operationalize the principle, in theory. This formal framework also lets us go from theory to working model. Working models attempt to link theory to real-world application by implementing formal specifications of the theory, usually with the help of computer programs. They are built not only for practical application, but also to allow us to experiment with the theory, thereby improving it. They also help make the theory more understandable, by showing its inner workings.

To implement a working model of precaution, we can build a computer program known as an *expert system*. Very simply, expert systems use computer programming to try to emulate the way real-world experts think about a problem. Our formalization of the precautionary rule of thumb in terms of fuzzy membership functions can be used to construct a working expert system.

The idea behind the fuzzy interpretation of precaution is straightforward. The math required for implementation is fairly simple. The challenge lies in identifying fuzzy probabilities. We could simply "draw" suitable membership functions for probability, based on expert opinions. That, however, would not get us deep enough into the expert thought process to qualify as a true expert system. We need to be able to somehow develop probability estimates from raw information, or inputs, about an action or event. Once the expert system program determines the probability estimate, we can apply rule of thumb. The result is a completely automatic process.

A.5 Artificial neural networks (ANNs) and human reasoning

Experts gain knowledge through experience. We need a process that will permit us to link raw data to expert opinion. One way of doing so is using artificial neural networks (ANNs). ANNs are computer programs that emulate the working of neurons in the human brain to "train" a model based on input data. While they sound complicated, they are fairly easily implemented in a variety of standard programming environments, including the ubiquitous desktop spreadsheet program (e.g., Microsoft^{tm} Excel).

ANNs are formulas for advanced calculation that are based on the functioning of neurons in the human brain. A neural network consists of various

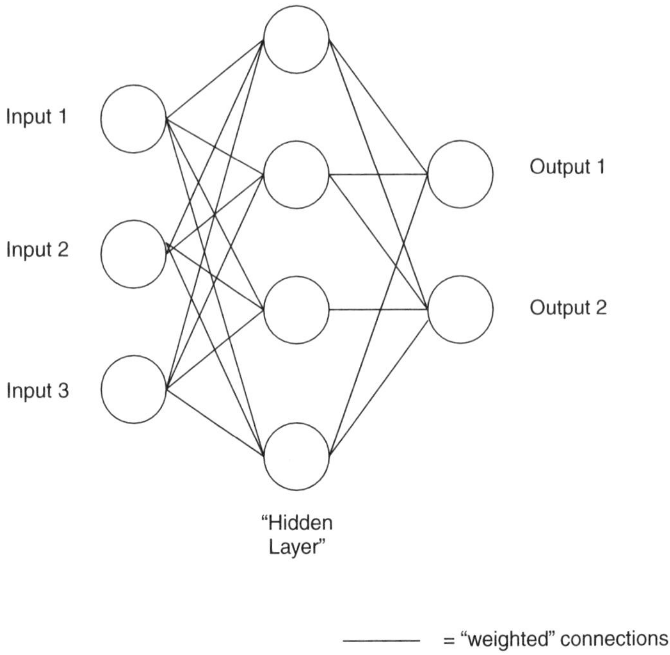

Figure A.4 A Simple Neural Network Architecture

processing elements (called nodes) than can be interconnected. Each note is "triggered" to various degrees, defining pathways from "input" to "output". These pathways are defined by different mathematical weights given to the data that goes through them. The nodes in the network are activated to various degrees, according to mathematical weights assigned to the interconnecting pathways. It is this process of differential activation of nodes that very efficiently captures the relationship between input and output.

Figure A.4 shows a very simple neural network architecture that relies on three layers of nodes, three in the first (input) layer, four in the second (hidden) layer, and two in the third (output) layer. The number of nodes, and the number of layers of nodes are determined by various rules of thumb. For example, the number of nodes in the "hidden" layer should exceed the number of inputs. More complex structures are therefore needed to represent complex relationships (i.e., high number of inputs and outputs).

Neural networks are easily imbedded in computer programs. By processing data in parallel ("all at once") as rather than serial ("one at a time") fashion, common to traditional computer programs, ANNs can process enormous amounts of data very quickly. This makes it practical to seek complex relationships in data by sifting through many possible combinations. This "data mining" process, known as training, is speeded up using optimization techniques that search for the most promising weighting scheme to represent relationships among data.

ANNs train on input-output data presented to them, in an attempt to make the error between network predicted output and actual output as low as possible. The desired level of "fit" is chosen by the decision-maker. It may depend on the problem at hand, and the availability of resources (running large networks many times takes time and computing resources).

A remarkable feature of well-trained ANNs is their ability to interpolate or fill in relationships among missing data. This means we can use a trained ANN to relate outputs to novel input data beyond (but not too far beyond) that used for training. This generalization capability also allows us to deal with more realistic continuous data, rather than a few select landmarks.

We will use a simple, but powerful, ANN structure to develop fuzzy probability membership functions, such as those shown in Figure A.3. We can then match these probabilities to our fuzzy definition of "possibility", and take precautionary action accordingly. For illustration, we will use a simplified model of dam safety. The model will be based on three simple inputs: age of the dam, its condition and its type. Age is input as years since "first fill". Condition is rated subjectively on a scale of 1 (poor) to 5 (good). The types of dams examined are either earthen or concrete. Factors and data are reasonable, but hypothetical. They are meant simply to illustrate how a practical model would work.

The model output is shown in terms of a fuzzy probability of catastrophic dam failure. Fuzzy probability membership functions are defined mathematically using a simple functional form know as the Gaussian. A Gaussian membership function is defined by three parameters. One indicates the center of the membership function (i.e., "best guess" probability of loss), another represents the "spread", or width of the function, and the third varies the "shape" of the function. To simplify, we will hold the shape parameter fixed, and only vary the center and spread. The architecture of this simple model is the same as the one shown in Figure A.4.

Modeling via ANNs begins with a set of "training examples". These examples are generated by asking experts to link various sample inputs (age, condition, type) to fuzzy failure probability graphs. The graphical output is then translated into the two function parameters. So, the training examples here would consist of a column of three numbers (inputs) related to a column of two numbers (outputs). For example, our expert (s) may link a 90 year old earthen dam in poor condition ("2") to a relatively certain (low spread) membership function around some fairly high probability of loss number, say one in a thousand, or 10^{-3}. On the other hand, a 20 year old, well maintained ("5") dam may be linked to a fairly precise estimate of low failure probability, resulting in the training examples. Situations of uncertainty may arise. For example, we may have a fairly new (25 years old) +earthen dam, in poor condition ("1"). Uncertainty arises in that it is unusual to find newer dams that are (a) earthen and (b) in such poor condition. Our input data would be linked to a fuzzy membership function for probability of failure that is fairly wide.

The ANN program then "trains" based on the data, using standard computational algorithms. For optimal training, a rather large set of training examples is required (say, 100 to 250, perhaps more, depending on the number of inputs). Training usually involves multiple runs (usually, hundreds) through the data. Parameters of the ANN model are modified on each run, according to the direction (+ or –) and rate of change ("gradient") of the measured difference, or error, between model output and the expert input/output combinations (the training examples). Training stops when the error hits some preset (low) level. The trained network is represented by a numerical matrix of "weights". These can be used in the expert system program to provide output for any novel inputs given.

Figure A.5 shows the output screen of our trained ANN expert system, implemented in Excel. ANN weights are stored, and calculations take place, in a "hidden" area of the worksheet. The user simply provides the input data. The example shown is for a relatively new concrete dam, in poor condition. As it is unusual to find such a new dam in such poor condition, considerable uncertainty surrounds the estimate. While new dams are usually relatively "safe", the condition issue suggests at least the possibility of a higher failure probability.

The fuzzy probability determined by the ANN is superimposed on a possibility ("risk") membership function, like the one shown in Figure A.2. We can take precautionary action based on a variety of criteria. Here, we use the degree of overlap between measured probability and possibility. The threshold degree (alpha level) in this example is set at 25 percent overlap. As this dam shows a probability whose uncertainty overlaps the risk threshold to a degree greater than 25 percent, the action indicated is "Use Precaution". In other words, we need to take immediate action to eliminate the possibility of dam failure, either *via* suitable restoration or replacement. In this way, the risk assessment/risk management process has been, essentially, automated.

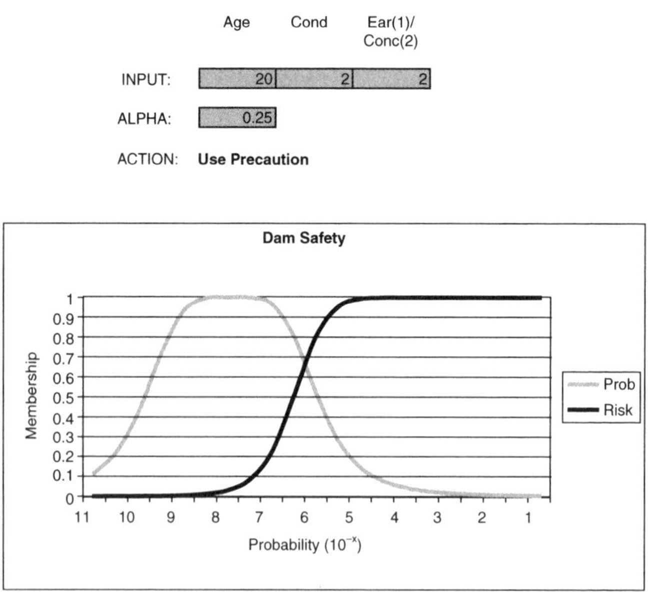

Figure A.5 A ANN-Based Fuzzy Risk Assessment System

The simple ANN based implementation of a working model for precaution provides a platform for investigating real-world implementations of precautionary risk management. Interesting extensions can be developed, such as the remote monitoring of risk based on direct electronic data acquisition (EDA) of relevant inputs. In our dam safety example, we could use some sort of remote condition monitoring in conjunction with the computer systems real-time clock, remotely linked to our computer. This would permit the automatic triggering of precautionary activities, such as the operation of spill gates. Action can be taken at time when the human decision-maker would be most stressed (i.e., in emergencies).

A.6 Using the model

Fuzzy set theory lets us formalize the uncertainty inherent in linguistic articulations of the precautionary principle for dealing with high-stakes risk. In doing so, it allows us to preserve the natural uncertainty inherent in rules about risk, and related measurements. This uncertainty comes not only from randomness, but also results from knowledge imperfections surrounding complex and dynamic systems. We don't know how to specify a lot of things about high-stakes risk exactly, so we use approximations: When the *possibility* of catastrophe exists, we avoid the action or event. The linguistic term "possibility" can in turn be defined using fuzzy sets.

The fuzzy formalization allows us to build a working model of precaution, using expert systems based on artificial neural networks. ANNs allow us to capture human expertise about high-risk situations using computer programs. A simple, but realistic, working model of precaution is presented here. Implemented in the easy to use, and widely available spreadsheet program Excel, the model shows how the risk assessment process can be automated while preserving inherent uncertainties.

The development of a formal model of precaution based on fuzzy sets also lets us focus on fruitful areas of further research. Among them, is the identification of membership functions associated with various exposures. How does science inform the process of membership function creation? Dynamically, how is this science used to narrow, or widen, the breath of these functions (as a representation of uncertainty)? Much membership construction is currently based on the input of human experts. We need to know more about how these perceptions are formulated in the minds of the experts. Inductive (experimental) associations, such as those captured by artificial neural networks are one way. Explicit modeling of the process of catastrophic loss, via qualitative event trees and the like, is another.

Our definition of "possibility" and our threshold for risk is another area for study. How do we define a proper level of uncertainty for this expression? Here, and throughout the discussion of precaution, we have assumed the threshold remains the same for all exposures. Might we capture ideas like the difference between natural and man-made perils within possible differences in threshold between the two? The proper specification of this threshold, in terms of fuzzy sets, is also important to its proper application to the real-world.

The appropriateness and workability of precaution can only be determined when we have an accurate model for the process. This does not mean the model has to be one that can be completely specified in mathematical terms. In fact, we have shown that the desire for too much precision can be counterproductive. Such accuracy is simply not available in the real-world. The model must accommodate these features of the uncertain world. We develop the model, work with it to provide the best representation of the theories that underlie it, and see if it works. If not, we go on to the next idea. The fuzzy set model of precaution in this way represents a true "working hypothesis" on the structure of high-stakes decisions. To the extent it, or its variants, are validated through application, we end up with a valuable tool for high-stakes decision-making.

Appendix B The Precautionary Risk Manager's Bookshelf

What follows is a bibliography of sorts, including references to works pertinent to topics discussed above, as well as more general reading on the "precautionary principle". As mentioned above, a considerable amount of risk management proceeds on the basis of statistical analysis of risk. There are literally hundreds of books on this topic. Most risk management textbooks focus on the statistical approach, many exclusively. While the management of statistical losses is important, these losses constitute the mere tip of the iceberg of all risk we face. Their disproportionate number with respect to the impact of risks is probably due to the fact that, while they can be quite complex theoretically, practically, they avoid many of the hard issues we face with high-stakes criteria. If nothing else, they emphasize our penchant for deceiving ourselves when it comes to the truly tough issues of risk in this world.

There exist some books and articles that purportedly deal with "catastrophic" and high-stakes risk management. Most of these simply extrapolate statistical techniques into the low probability/high consequence domain. As we have seen, such applications are useless at best, and can actually be quite misleading. Statistical applications in high-stakes risk are a fiction. Statistical methods do not apply to catastrophic risks (as we have defined them). Armed with the knowledge supplied above, the astute reader will be able to spot those books that purportedly deal with high-stakes risk, yet use inappropriate statistical methods.

Most books and articles cited here take an explicitly precautionary approach to high-stakes risk. Most of these are also critical of the simplistic application of statistical methods and expected value decision criteria to high-stakes risk as well. They represent both deeper and broader reviews of the challenges associated with catastrophic risk management, and the suggested solutions.

The reader will also notice that precautionary efforts are most often suggested, and discussed, in connection with global challenges of an extremely widespread dimension, including environmental issues such as global warming and the use and disposal of carcinogenic chemicals. While these "grand challenges" are an obvious focus of precaution, the precautionary lessons learned can be applied to the individual level, the business, community, as well as the wider society and global community. We have tried to express this in our approach above. The readers challenge is to take these ideas and apply them to his or her own domain.

We also cite here some books and articles that, while not directly about risk and risk management, provide supporting ideas for the precautionary risk management effort. For example, an understanding of uncertainty beyond simple randomness is necessary to be able to truly understand high-stakes risks and their challenges.

Remarkably, little has been written about what we have called the "fatalistic" approach to high-stakes risk: Do nothing (and hope for the best). This despite the fact that we have noted that it may in fact be the dominant approach to high-stakes risk today. While there is little guidance to be given on doing (not doing?) fatalistic risk management, the implications for the treatment of risk are huge. And in fact there may be implications associated with fatalism that the fatalist may ignore. We have suggested that the position may be "taken advantage of" by those that would seek to further self-interests. There are probably many more aspects of the fatalistic approach that have not been fully developed, even by devout fatalists. Undoubtedly, the lack of detailed study of the position comes from what might be a general distaste for inactivity in the face of adversity. The fact is, fatalism is a bona fide approach to high-stakes risk, and it is, consciously or unconsciously, widely practiced. It's certainly an area deserving of further study.

The level of specialized expertise required by these various citations varies. Most, however, will be accessible to the educated layperson. Precaution is a very intuitive approach to risk. It avoids many of the theoretical trappings of statistical risk analysis. That said, it behooves the reader to have a working understanding of statistical methods and probability theory in general. Only with this understanding is the risk manager "fully rounded" in his or her approach to risk. Nonetheless, the emphasis, as stated above is on precautionary treatments.

As the world gets more and more complex, issues of risk come to the forefront. This increasing complexity, and the perception of associated high-stakes risks it brings with it, will undoubtedly result in more and more studies of high-stakes decision criteria. The reader is advised to be on the lookout for literature that expands our understanding of catastrophic risks and their treatment.

The selected bibliography

Ackerman, F. and R. Massey, *Prospering With Precaution: Employment, Economics and the Precautionary Principle*, Global Development and Environment Institute, Tufts University, August 2002. Suggests that the fear of precaution on an economic basis is exaggerated (except perhaps by those that have most to fear from the principle). This study was commissioned by the Alliance for a Healthy Tomorrow, and is available on their website.

Arrow, K. and L. Hurwicz, "An Optimality Criterion for Decision-Making Under Uncertainty", in Carter, C. F. and J. L. Ford (eds) *Uncertainty and Expectations in Economics*, Augustus M. Kelly Publishers, 1972. An absolute mass of complicated mathematical equations after the first page, this paper will take an advanced degree in economics to understand. It is, however, often cited as a strong theoretical argument that ONLY the minimax (risk avoidance) solution encompassed in precautionary risk management, or it direct opposite, fatalism (i.e., risk acceptance) make sense when dealing with high-stakes risk when probabilities are unknown (or irrelevant). There is no "in between" (e.g., expected value). Based on a widely circulated, yet unpublished 1951 paper by Hurwicz, entitled "Optimality Criteria for Decision-Making Under Ignorance" (Cowles Commission Discussion Paper: Statistics, No. 370, December, 1951). For what its worth, Kenneth Arrow is a Noble Prize winning economist. Hurwicz is a "founding father" of modern decision theory.

Bankes, S., "Exploratory Modeling for Policy Analysis", *Operations Research*, May–June 1993. Early, and still one of the best, papers on what exploratory modeling is, and how to apply it to "assist in reasoning about systems were there is significant uncertainty". Consolidation of facts into models based on "average", or "best guess" behavior was a matter not so much as our "proper" view of science, as the availability of analytical tools which limited our ability to model all the possibilities. This barrier was essentially broken with the increased availability of powerful computers.

David, F. N., *Games, Gods and Gambling*, Hafner, 1962. Describes the origins of formal thinking about probability, up to what we have called the first probability "revolution" in the late 17th and early 18th century. Much of this early formal development was based on analysis of games of chance. Unfortunately, the precision with which we could specify the workings of simple gambling devices did not translate well to complexity of the real-world. As a result making probability "the very guide to life" (as Bishop Butler suggested) was not quite as easy as it seemed. Despite the tremendous progress in the sciences made through the application of probability theory, it was not until the late 20th century, and the recognition of the significant role that knowledge imperfections made in the application of probability, that probability found its proper place among methods of describing generalized indeterminacies. See the book by Smithson, cited below.

Funtowicz, S. and J. Ravetz, *Uncertainty and Quality in Science for Public Policy*, Kluwer Academic, 1990. All science is not created equal.... The authors attempt to establish criteria for "good" science, and find that all science does not uniformly fit those criteria. They suggest that scientific results be accompanied by a sort of scorecard for measuring its reliability, in the face of complications presented by complexity and dynamics.

Haimes, Y. Y., *Risk Modeling, Analysis and Management*, Wiley-Interscience, 2004. While this book focuses on statistical methods of dealing with risk (it is one of the best on this area), it also recognizes their limitations. In Chapter 8, Dr. Haimes addresses what he calls "The Fallacy of the Expected Value", citing many of the arguments we have developed here. His own solution to the problem of high-stakes risks involves the extension of expected value known as conditional expected value. Conditional expected values are calculated based on partitions of the probability distribution of loss, thereby giving greater emphasis to the low-probability/high-stakes end of the distribution of risks. While supporting more conservative risk management decisions, the approach does not fully escape the difficulties of applying "short-run" solutions to "long-run" problems.

Haller, S., *Apocalypse Soon? Wagering on Warnings of Global Catastrophe*, McGill-Queens University Press, 2002. Perhaps the best overall review of the underlying philosophy of high-stakes risk assessment and decision-making since Rescher's *Risk: A Philosophical Introduction ...* (cited below). This book shows, once again, why expected value/cost-benefit decision-making doesn't work for high-stakes risk, and presents some compelling arguments for precaution.

Herremoes, P., J. Keys, M. Macgarvin, A. Stirling and Vaz, S. G. (eds), *The Precautionary Principle in the 20th Century: Late Lessons from Early Warnings,*

Earthscan/James & James, 2002. Lessons learned (?) from failure to apply precaution. Examines asbestos, CFCs and ozone depletion, PCBs, pollution in the Great Lakes and other public disasters. Preventable by precaution? You decide. Based on these case studies, the authors offer some insights into the effective use of precautionary strategies to prevent future disasters.

Jablonowski, M., "Managing the 'Iceberg' Model of Risk", *CPCU Journal (Society of Chartered Property and Casualty Underwriters)*, July 2005. Statistical risk remains the mere "tip of the iceberg", with catastrophic loss potentials looming large, and occasionally surfacing (with disastrous results). To manage "what lies below", we need to pay attention to the specialized methods of managing high-stakes risk.

Jablonowski, M., "Facing Risk in the 21st Century", *Risk Management*, June 2004. Where does precaution "fit in" to risk management strategies for the 21st century? Suggests that "fatalism" (do nothing, and hope for the best) is the dominant risk philosophy today (like it or not).

Jablonowski, M., "Automating the Risk Assessment Process", *CPCU Journal (Society of Chartered Property and Casualty Underwriters)*, Summer 1998. Introduced the idea of automating risk assessment using fuzzy logic and neural networks. Good references on the basics of artificial neural networks, and how they work.

Jablonowski, M., "A New Perspective on Risk", *CPCU Journal* (Society of Chartered Property and Casualty Underwriters), Winter 1996. The concept of risk defined using fuzzy sets. Suggests how the spectrum of risk is treated by organizational risk managers, using the concept of the *Fuzzy Risk Profile*.

Kosko, Bart, *Fuzzy Thinking: The New Science of Fuzzy Logic* (Hyperion, 1994) A good, non-technical introduction to fuzzy sets and their related logic. Also discusses the use of the use of fuzzy sets to operationalize linguistic "rules of thumb".

Lowry, R., *The Architecture of Chance*, Oxford University Press, 1989. The focus of this book is on probability theory and statistical risk. Hundreds of references on the statistical treatment of risk exist. This little book remains one of the best and most complete treatments in this domain. Should be read by all who want to complete their understanding of risk in all facets.

Luce, R. D. and H. Raiffa, *Games and Decisions: Introduction and Critical Survey* (Wiley, 1957). A classic in the theory of decision-making. Covers decisions when the probabilities of loss are known ("risk") and unknown ("uncertainty"). Emphasis, however, is decidedly on risk and related statistical methods.

Morris, J. (ed.), *Rethinking Risk and the Precautionary Principle*, Elsevier Science & Technology Books, 2000. Precaution is not without its pitfalls. As we have noted, when consistently applied, it is "expensive" (both directly, and in terms of "opportunity costs"). As it is modern industry that usually bears these initial costs, it is not surprising that some business interests are "up in arms" over precautionary ideas in risk management. Books like these put forward the anti-precautionary arguments, often quite vehemently. The arguments would be far more convincing if they were not made by researchers and scholars who are funded by those that create the risks in the first place.

O'Brien, M., *Making Better Environmental Decisions: An Alternative to Risk Assessment*, MIT Press, 2000. An "alternatives analysis" framework for implementing precautionary risk management. Suggests that progress and precaution are not mutually exclusive.

Raffensberger, C. J., J. Tickner and W. Jackson (eds), Island Press, 1999. *Protecting Public Health and the Environment: Implementing the Precautionary Principle*. Edited volume of conference proceedings. Very accessible, self-contained introduction to precautionary principles. Like many books on this list, the emphasis is on community risk management of environmental and public health issues. This is the area were many of our toughest risk challenges lie. All the ideas here are applicable to other high-stakes risks, whether they are viewed from an individual, organizational or societal perspective. Of benefit to risk managers, at whatever level they serve.

Rescher, Nicholas, *Risk: A Philosophical Introduction to the Theory of Risk Evaluation and Management*, University Press of America, 1984. An eminent scientific philosopher looks at risk from the wider perspective. Excellent coverage of the treatment of "high-stakes" risk, and related complications. Rescher's book *Luck* is a worthy companion.

O'Riordan, T. and J. Cameron, *Interpreting the Precautionary Principle*, Earthscan/James & James, 1994. A collection of essays that appeared early on in the "formalization" of precaution as a principle of catastrophic risk management.

Perrow, C., *Normal Accidents: Living With High Risk Technologies*, Princeton University Press, 1999. Shows how new technologies can bring with them greater risks, simply by virtue of their (often unfathomable) complexities. Cities and reviews in depth several real-world examples.

Schrader-Frechette, K., *Burying Uncertainty: Risk and the Case Against Geological Disposal of Nuclear Waste*, University of California Press, 1993. A case study which points out the many inconsistencies and inaccuracies that result in applying high-stakes risk assessment methods based on statistics and expected values. The author concludes that many of these efforts are driven by self-interest rather than their purported "scientific" basis. Along with her book *Risk and Rationality* (cited below), these writings represent some of the deepest thinking on the subject of high-stakes risks.

Schrader-Frechette, K., *Risk and Rationality*. Argues that pseudo-scientific risk analysis based on statistical risk assessment are often used to further hidden agendas. Also, suggests that the public's seemingly "irrational" fears and concerns over man-made risks with potentially catastrophic exposures, and their general mistrust of "scientific" assessments based on statistics and expected values, are often quite rational. In Chapter 8, Schrader-Fechette argues the pros and cons of an approach similar to the minimax (with respect to catastrophic losses). Ultimately, however, her recommendations for a rational risk policy hinge on what she calls "scientific proceduralism" in which the analysis of probabilities and consequences, and even expected cost/benefit, is retained. Proceduralism requires that the risk assessment process be made more open, and hence more "democratic". Undoubtedly, this approach would lead to more honest probability and consequence assessments. At issue is if any such assessments can overcome the catastrophe problem: Do any probabilities (except the smallest) make sense when faced with extinction? Scientific proceduralism is reviewed, and critiqued, at length in the book by Stephen Haller, cited above. A real in-depth analysis of many issues surrounding high-stakes decision approaches that gets pretty dense at times. Some of the other books suggested here are better for a more concise introduction and overview of some of the pertinent issues.

Sjoberg, L., "Three Themes in Risk Perception: Toward a Conception of Risk Perception as Ideology Rather Than Emotion". In L. H. J. Goosens (ed.), *Risk Analysis: Facing the New Millennium*, pp. 369–373, Delft University Press, 1999. Lennart Sjoberg, Director of the Stockholm School of Economics Center for Risk Research, reviews three major dimensions in the theory of risk perception: The primacy of risk over benefits (the "hegemony of risk"), tampering with nature, and trust. Sjoberg, more than any other researcher in this field, promoted the idea that risk is *not* "all in our heads".

Smithson, M., *Ignorance and Uncertainty*, Springer-Verlag, 1989. Provides a taxonomy and survey of the various types and great variety of uncertainties we face in life. Details what we have called the second revolution in uncertainty: The recognition that other forms of uncertainty exist beside randomness, as specified in the formal theory of probability. The recognition that others forms of uncertainty exist, and indeed co-exist, must be recognized if we are to make effective decisions in the modern world.

Sunstein, C. R., *Laws of Fear: Beyond the Precautionary Principle*, Cambridge University Press, 2005. Professor Sunstein mounts the "standard" attack on precaution (too expensive, makes "everything" risky, leads to a paralysis of fear, etc.). Suggests further that precaution attracts so much attention today because of peoples subjective "amplification" of risk (apparently, it's all in our heads). The author further cautions that precaution must not be applied to everything (of course not!). Some (smaller) risks can be effectively managed using the statistical approach). Ultimately, the author suggests the use of an "anti-catastrophe" principle that ends up looking a lot like precaution. In the end, he does not get us all that far beyond the precautionary principle. This book illustrates why a more formal definition of precaution, using ideas such as those demonstrated in Appendix A, are a prerequisite for rational discussion of precautions "pros and cons". An influential and articulate author, Mr. Sunstein is a law professor at the University of Chicago.

Taylor, R., *Metaphysics*, Prentice Hall, 1991. Contains a chapter on the philosophical argument for fatalism. While the fatalistic approach to our existence is much broader than what we have been calling fatalism with respect to risk, they arguments have similarities. While Taylor's fatalist would clearly not take precautions with respect to risk, they would equally apply the fatalistic approach to all aspects of life. The idea seems a bit extreme at first reading. Nonetheless, Taylor makes some strong arguments for this generalized "existential" fatalism, and further suggests people may find comfort in such an approach. Our review of risk acceptance criteria suggest that people may find a similar comfort in fatalism with respect to what we have called "natural" risks.

Tickner, J. (ed.), *Precaution, Environmental Science and Preventive Public Policy*, Island Press, 2003. Emphasizes the interface between science and precaution. Includes sections on scientist's perspectives on precaution; precaution, ethics and the philosophy of science; uncertainty in science and public policy (including case studies); science in governance; and science to support precautionary decision-making. All essays here are very readable, and contain many additional references for the interested reader to pursue. Joel Tickner is currently an assistant professor at the University of Massachusetts at Lowell, in the School of Health and Environment's Department of Work Environment. He has applied precaution to real-world problems, many in an industrial environment. His research interests in this area include the study of long-term chemical policies, and methods of clean production.

Zimmermann, H.-J. *Fuzzy Set Theory and Its Applications* (Kluwer Academic Publishers, 1991). A more technical introduction to fuzzy set theory, with applications.

Appendix C A Concise Statement of the "Precautionary Principle"

Where the threat of serious or irreversible harm to people, property or the environment exists, actions must be taken to remove this threat. These actions include prudent avoidance, loss prevention efforts, and the search for effective alternatives. Lack of full scientific certainty as to cause and effect shall not preclude these preventive actions when at least the possibility *of catastrophic impacts exists.*

Appendix D A Glossary of High-Stakes Risk Management

Alternatives Assessment The application, early on in planning for progress, of a search for alternatives that may eliminate future potentials for catastrophe. By applying alternatives analysis early, and with an eye to the dynamic character of risk, the **dilemma of precaution** may be prevented.

Average A property of events that occur randomly that allows us to determine useful parameters of the outcome over time. Averages can be substituted for deterministic values in optimization when we can observe a sufficiently long sequence of outcomes.

Avoidance Avoiding risk, either by not engaging in an activity, or by instituting effective preventive measures, is at the root of **precaution**.

Burden of Proof With regard to high-stakes risk, suggests that anyone who proposes activities (especially for profit) be responsible for proving their safety (i.e., that there is no reasonable **possibility** of catastrophe associated).

(The) Catastrophe Problem When dealing with **catastrophe**, we don't get a "second chance". This greatly complicates the decision process. The **average** or **expected value** of loss is of no use here, because in the long-run, the long-run may cease to exist.

Catastrophic Risk Random losses that entail serious, perhaps irreversible consequences, such as the fatality of the individual, business ruin, or a threat to the continuation of humankind in general.

Confidence Intervals A statistical device used to express error due to sampling. Confidence intervals are determined by applying the mathematical theory of probability. The express uncertainty due to **randomness** and should not be confused with simple **intervals** based on possibility (see **knowledge imperfection**).

Cost/Benefit Analysis A simple optimization procedure, based on comparing the resource expenditures and/or opportunities lost ("costs") against rewards ("benefits").

Decision Matrix (Table) Specifies "actions" (the things we can control) on one axis, and "states of the world" (the things we can't control) on the other. Outcomes are shown at the intersection of each combination of actions and states. Useful for showing the effects of risk management options (actions) on potential losses (states of the world).

(The) Danger Zone See **Precautionary Region**.

(The) Dilemma of Precaution The **minimax** basis of **precaution**, consistently applied, means we should be willing to spend up to the amount of loss to prevent it. In high-stakes situations, the high-cost of precaution could itself have catastrophic consequences. We in effect become "doomed if we do, doomed if we don't". Prudent application of alternatives assessment early on in the process of determining the best path for progress could help eliminate such dilemmas.

Disutility A subjective measure of the loss we experience, calibrated in term of monetary units, or other possible objective loss measures (i.e., fatalities). Where disutility shows a one-for-one relationship with objective loss measures, we say the decision-maker is *risk neutral*. Where disutility is higher than objective loss we say the decision-maker is *risk averse*. Risk averse disutility functions suggest that decision-makers value larger losses disproportionately higher than smaller ones. Disutility, however, remains a very subjective measure of risk aversion.

Event Tree A logic diagram that breaks the process of loss development down into its component parts. From initiating event, the potential paths of a loss, along with their associated probabilities and consequences, are following to logical outcomes along the "branches" of the tree. Probabilities of various outcomes can be computed from the probabilities of various sub-events. In this way, we can infer the probabilities of more infrequent events without the need for statistical data. Failure to specify the probabilities of sub-events correctly, or the logic of the tree, can result in inaccuracies. Therefore, event trees should be used as more of a mere *approximation* to high-stakes probabilities and potentials. Qualitative events trees rely less on the exact specification of probabilities and can provide valuable information on catastrophic loss potentials.

Expected Value A generalization of the concept of average determined by multiplying the probability of an event by its outcome. For example, the expected value (loss) of an auto accident costing $10,000 with a probability of .1 is $1,000 (.1 × $10,000). Over a sufficient number of repetitions, actual outcomes will equal the expected value. Expected values measure the "benefit" component of the cost/benefit optimization of risk treatment options. If the expected value of a loss exceeds the cost of prevention, we utilize prevention, and *vice versa*.

Exploratory Modeling The specification of multiple plausible models based on uncertain inputs. Decisions based on exploratory models therefore take into consideration an ensemble of possible models. Exploratory modeling takes into consideration knowledge imperfections in modeling the complex real-world. Unduly precise models, based on consolidation of possibilities, can mask this uncertainty, resulting in misguided decisions. Exploratory modeling is very much a process of discovery, of what we don't know, as well as what we do.

Fatalism An approach to decisions that suggests that when we can't do anything about risk, we don't. The fatalists credo is, Why worry? Also sometimes referred to as **risk acceptance**. While a fatalistic, or "do nothing" attitude toward loss often carries a negative connotation, fatalism does force us to recognize our immediate powerlessness in the face of high-stakes risk. It is certainly better than fooling ourselves. In fact, a fatalistic (**minimin**) view of risk, along with the alternative of **precaution** (based on **minimax**) can be shown to be the *only* two logical alternatives when loss probabilities are unknown (or irrelevant).

Fuzzy Sets A characterization of concepts that are imprecise due to knowledge imperfections, that allows us to grade our degree of confidence in a concept belonging to a certain set of concepts. By allowing for knowledge imperfections, fuzzy sets and their related logic let us more realistically deal with a dynamic and complex world. Fuzzy sets are a generalization of simple intervals.

Hazard Risk This form of risk arises from physical properties of the world, and manifests its affects directly on *nature*. Nature includes people and their property. This is in contrast to the study of economic risks, which affect primarily financial variables. Hazard risks can themselves be classified as natural and man-made. Natural risks include the age-old perils of fire, wind, earthquake and flood. The modern world introduces a variety of perils produced by man. These include the risks associated with technological progress, including the failure of technology to function as intended, resulting in physical harm to humans and their environment.

Hurwicz Criteria A decision criteria that attempts to reconcile fatalism (minimin) and precaution (maximin) using an index of optimism/pessimism (α) to blend the two. A problem with the Hurwicz Criteria is that the choice of α can only be justified subjectively.

Iceberg Model of Risk Suggests that, like with an iceberg at sea, serious loss potentials lie "beneath the surface" of observable, **statistical risk**.

Insurance A financial mechanism whereby the effects of a single large loss on an insured is reduced due to a sharing of losses among the wider pool of policy holders. Clearly, the mechanism can be overwhelmed by losses that are sufficiently large so as to overwhelm (ruin) the pool.

Identify-Assess-Treat (I-A-T) Model A model for risk management that suggests the process proceeds by identifying exposures to risk, assessing their potentials in terms of probability and consequences, and choosing the best alternate treatment of that risk. Most suited to statistical risk, due to the ability to "self-correct" in such situations. When applied to catastrophe risk, the I-A-T model can result in a reactive stance toward risk, that may mislead risk management efforts.

Interval A range that encompasses the "true", but unknown, value or measurement of some property. Intervals are used to specify **knowledge imperfection**. In fact, the width of an interval can provide a measure of this type of uncertainty.

Interval Expected Value An **expected value** calculation based on an **interval probability** instead of a single probability point. The interval expected value lets us capture uncertainties due to knowledge imperfection, but since it still represents an average measure, does not solve the **catastrophe problem**.

Interval Probability Using an **interval** instead of precise point estimates to specify probabilities of events that are subject to **knowledge imperfection**.

Knowledge Imperfection A dynamic and complex world means that we can know many facets of this world only imperfectly. Knowledge imperfection is a form of uncertainty that is measured in terms of possibility. Possibility, in turn, can be specified using **intervals** and **fuzzy sets**. Knowledge imperfection as a form of uncertainty is distinct from **randomness**, though the two forms often coexist (see **interval probability**).

(The) Local Fallacy When dealing with catastrophic risk, precaution must be applied globally, that is, to all catastrophic risk, not just a "chosen few". **The**

catastrophe problem operates on a wide scale. A global view of precaution exacerbates the **dilemma of precaution**.

Minimax A decision criteria which suggests we choose risk management strategies so as to minimize the maximum loss, regardless of probability. Minimax lies at the root of **precaution**.

Minimin A decision criteria that characterizes **fatalism**: Minimize the minimum loss.

Null Hypothesis In statistical decision, the hypothesis of "no effect". In dealing with catastrophic loss potentials, we may frame the null hypothesis as "Activity/ event x has no catastrophic effects". See **Type I Error** and **Type II Error**.

Opportunity Cost The cost of forgone benefits.

Optimization The process of seeking the best alternative, in terms of some objective, or goal.

Possibility As used in conjunction with precaution, suggests some minimum level of potentiality below which we believe, for all intents and purposes, the event can not occur (or, at least, we should not be concerned with it). The threshold for possibility can be set in terms of very low probability, or perhaps based on physical considerations of complexity. Some non-zero threshold of possibility is needed if **precaution (minimax)** is to become workable. Otherwise, *everything* becomes "possible", and hence *everything* should be avoided.

Post-Fact Risk Management Management of risk after the process causing the risk has become entrenched. Examples are treating pollution threats via end-of-pipe treatments and fines is post-process, and relying on product liability insurance in the event of product related injuries is post-market. Post-fact methods may work in an environment of **statistical risk**, where errors can be systematically identified and corrected. Post-fact risk management is not appropriate for catastrophic risk, and can lead to the subsequent **dilemma of precaution**. Catastrophic risk requires pre-planning in the form of **alternatives assessment**.

Practicality See **proportionality principle**.

Precaution As used in this book, precaution refers to an approach to high-stakes risk based on the **minimax** principle of decision-making under uncertainty. Most simply, precaution suggests we avoid catastrophic risks. To make precaution workable, we need to also specify some non-zero threshold for **possibility** of risk as well.

Precautionary Principle A term used to describe linguistic articulations of precaution.

Precautionary Region That region of probability loss space that suggests that activities or events falling into this region should be avoided. Known more colloquially as the "Danger Zone".

Precautionary Regulation Public policy for the control of activities based on the principle of precaution.

Probabilistic Risk Assessment (PRA) The use of a variety of, mostly statistical, methods to determine the probabilities of loss, including those with catastrophic results. As the results of PRA are almost always in the form of precise estimates, uncertainty due to knowledge imperfections is often ignored. The result is unduly precise estimates, especially those pertaining to high-consequence/low probability events where data is scarce. Usually associated with **expected value** decision-making.

Probability When events are uncertain due to **randomness**, the best we can do is get some idea of their long-run relative frequency over time. We call this long-run frequency of occurrence *probability*. The probability of event x, p(x), can be defined as,

$$p_x = \frac{\text{Number of outcomes in which x occurs}}{\text{Total number of Outcomes}}$$

Proportionality Principle Suggests that **precaution** is subject to "reasonable" cost constraints. Precautionary actions restricted in this fashion are subject to the **catastrophe problem**.

Randomness A form of uncertainty that is due to our inability to completely specify the initial and subsequent conditions of some chain of events. Implies a variability in results that can not be controlled, i.e., selecting from a mixed bowl of colored balls blindfolded. In order for us to be able to specify randomness, at least the potential outcomes must be known.

Risk In general, anything involving the likelihood of adverse consequences. Risk can have multiple meanings, depending on the context, however. For example, risk is often used as a synonym for danger: Skydiving is *risky*. The various meanings must be carefully delineated in practical usage of the word.

Risk Acceptance See **fatalism**.

Risk Acquiescence Conditions under which risk is not so much accepted as forced. Results in a feeling of resignation to risk, and is a passive acceptance to risk, as opposed to a reasoned acceptance. To the extent acquiescence results in worry and fear, we can not say that the goal of **risk management** has been fulfilled.

Risk Aversion See **disutility**.

Risk Communication The process of informing affected parties about risk.

Risk Management The process of reducing our worry over the risks we face.

Risk Perception Factors affecting the reasoned acceptance of risk. Studies have shown that risk perception is often based on the naturalness of risk: The more natural a risk, the more acceptable. Naturalness is in turn reflected in factors such as fairness/equity, familiarity, degree of control, voluntary vs. involuntary and degree of trust. In this sense, risk perception is a mater of ideology rather than subjective feeling or emotion (i.e., risk is *not* "all in our heads").

Sampling The process of selecting a representative portion of a wider population on which to base inferences about randomness and probability. For example, what is the probability that a man has green eyes?

Sampling Error Random error that results from sampling from a population of random occurrences. Under controlled conditions, sampling error can be determined using mathematical formulas. Sampling error is expressed in terms of statistical **confidence intervals**.

Simulation Modeling of real-world systems based on expressions of their mathematical structure (i.e., equations), usually with the aid of a computer. Computer simulations let us observe the complex behavior of the modeled system in a controlled setting. Simulation also helps verify the underlying model against the real-world, by comparing the simulation results to actual observations. However, simulation in and of itself can not add to the credibility of the underlying model, as it is dependent on that underlying model for its results. In other words: Garbage in, garbage out.

Statistical Risk Risks that manifest themselves over a relatively observable time horizon (say 10, 20, maybe to 50 or 100 years or so), or whose conditions of occurrence can somehow be controlled (i.e., in product quality control applications), can effectively be treated by using averages and expected values and standard optimization techniques (e.g., **cost/benefit analysis**).

Strict Liability A legal doctrine that holds that those who would profit from the introduction of potentially catastrophic activities be responsible for any adverse results, regardless of "fault" or "negligence". Presumes that failure to take precautions in such cases establishes liability, prima fascia.

Ten-Percent Rule A rule-of-thumb that suggests that when the conditions of a statistical experiment can not be controlled, probabilities less than 10 percent (1/10, or .1) can not be accurately specified. While the rule is very conservative, it does appear that for things like large scale natural events (for example, earthquakes) precise estimates of annual probability (e.g., .001) do not appear credible, as they are difficult (if not impossible) to verify. As a result, they must be treated with caution when used in the decision process.

Type I ("Type One") Error Refers to statistical error of rejecting the "no effect", or null hypothesis, when it is in fact true. In high-stake risk analysis, we commit a type I error when we reject the hypothesis that an activity has no catastrophic effect, when in fact it *does*. Emphasis on reducing type I errors shows a bias toward progress, at the expense of possible catastrophe.

Type II ("Type Two") Error Refers to the statistical error of accepting the "no effect" or null hypothesis, when it is fact false. In high-stakes risk analysis, we commit a type II error when we accept the hypothesis that an activity has no catastrophic effect, when it in fact does. Emphasis on reducing type II errors shows a precautionary approach to catastrophic loss.

Uncertainty The inability to specify values exactly.

Urn Model of Probability A simple model that uses draws of colored balls from an urn, or bowl, to model the physical reality of probability.

Weighted Probability Probabilities may be weighted disproportionately in the decision process, indicating decision-makers subjective attitudes toward the size of the probability.

Index